Contemporary Trends and Issues in Science Education

Volume 51

The book series *Contemporary Trends and Issues in Science Education* provides a forum for innovative trends and issues impacting science education. Scholarship that focuses on advancing new visions, understanding, and is at the forefront of the field is found in this series. Authoritative works based on empirical research and/or conceptual theory from disciplines including historical, philosophical, psychological and sociological traditions are represented here. Our goal is to advance the field of science education by testing and pushing the prevailing sociocultural norms about teaching, learning, research and policy. Book proposals for this series may be submitted to the Publishing Editor: Claudia Acuna E-mail: Claudia.Acuna@springer.com

More information about this series at http://www.springer.com/series/6512

Valarie L. Akerson • Gayle A. Buck
Editors

Critical Questions in STEM Education

Springer

Editors
Valarie L. Akerson
Curriculum & Instruction
Indiana University
Bloomington, IN, USA

Gayle A. Buck
Curriculum & Instruction
Indiana University
Bloomington, IN, USA

ISSN 1878-0482 ISSN 1878-0784 (electronic)
Contemporary Trends and Issues in Science Education
ISBN 978-3-030-57645-5 ISBN 978-3-030-57646-2 (eBook)
https://doi.org/10.1007/978-3-030-57646-2

This Springer imprint is published by the registered company Springer Nature Switzerland AG
The registered company address is: Gewerbestrasse 11, 6330 Cham, Switzerland

Foreword to *Critical Questions in STEM Education*

For those working in STEM education as teachers, principals, teacher educators, and researchers, a central concern in recent years is developing a consensus on what STEM education can and should be, in terms of curricular content, pedagogy, and application to real-world problems. Perhaps heightening a sense of urgency regarding this task is STEM's near-juggernaut quality as an educational movement internationally. Meanwhile, a rush by various discipline advocates to claim curricular "terrain" in K-12 STEM has led to calls for STEAM (adding art), STREAM (reading), CSTEM (coding or computer science), and so on, which complicates development of a clear understanding of what STEM education should include. STEM as "ambiguous slogan" (Bybee 2013) nonetheless has rapidly diffused across many mass education systems, proving to be an effective tool to advocate for resources (Shaughnessy 2012). The contributions in this volume offer several cornerstones, comprising the parts of the book, from which to examine questions about the contours of STEM in a thoughtful and research-informed manner. The point of departure here is a working definition of STEM that includes a renewed focus on the variation across individual disciplines as well as the meaningful interdependence that connects disciplines constituting STEM.

Since the early days of STEM being promoted as a kind of curricular package, a frequent element of the sloganeering blithely portrayed STEM education as "integrated" and "interdisciplinary," even as curriculum scholars have emphasized the tremendous difficulty for interdisciplinary knowledge to secure a place in the school curriculum. STEM education scholars could benefit from prior work on the challenges of developing and implementing interdisciplinary curricula, however appealing their ring, such as in social studies and "humanities" (Ravitch 2003; Wineburg & Grossman 2000). In this volume, we find a serious attempt to conceptualize the limits of the interdisciplinarity of STEM, starting in the first part with a series of chapters articulating the "nature of" each of the four areas (extending Lederman's groundbreaking work on the nature of science) and their varied epistemological and ontological underpinnings. In an overview of this first part, Akerson and colleagues boldly suggest that given the substantial differences in the core natures of the disciplines (and even within each area), there can be no analogous and fully coherent "nature of STEM." If these scholars are right, the implicit question emerges regarding how truly integrated and interdisciplinary STEM can be.

This tension is illustrated in Part 2, which views STEM education from the ground up, considering approaches to teaching STEM, both at the level of the classroom and the school, but also the challenges in preparing teachers to support integrated STEM learning. The self-study by Yin (Chap. 7) is particularly illustrative on this point, as even a seasoned science teacher educator struggled to balance and integrate all four major fields in a STEM education course for pre-service teachers. University Technical Colleges in England (Dobrin, Chap. 8) offer an organizational form that affords opportunities and time to both integrate and apply STEM knowledge, but even there, students are encouraged to choose areas of particular interest to focus on during group projects (e.g., "Do the part you are interested in"), effectively de-integrating the STEM work to some extent.

The final part raises broader questions about perceptions of STEM by various stakeholders. Perhaps, in a sense, school-based STEM is what school STEM does. Newman and colleagues (Chap. 10) consider how schools certified as "STEM schools" by the state of Indiana portray STEM, while Sgro, Bobowski, and Oliveira (Chap. 11) systematically consider visions of STEM proffered by practitioner journals, demonstrating the difficulty of meaningfully integrating across all four areas. In both chapters, STEM integration is threatened by the dominance of one or more of the component disciplines. Sgro and his co-authors resolve this by taking the position that STEM cannot be a discipline in its own right, but rather should be seen as a "meta-discipline." When considering experiences and the STEM identity of college students majoring in and in some cases switching out of STEM, Song, et al. (Chap. 13) ground coding decisions about what is and what isn't a "STEM major" based on whether the major was located in the institution's College of Natural Sciences and Mathematics, which raises questions of how new or rapidly changing fields (like psychology) are classified with respect to the STEM umbrella. In the end, there are numerous echoes of the doubts raised in Part 1 about whether there can be a coherent "nature of STEM."

Rather than hunting down a perfectly balanced and interdisciplinary "quark" (Renyi, 2000) called STEM, the brightest potential for STEM education may lie in its core focus on engaging with complex, "ill-formed" problems, as highlighted in many of the contributions here. Comprising a vigorous pedagogical culture (Weld, 2017), rather than a strictly delineated and official school subject, the varied tools of STEM could be used as a springboard into learning to analyze Shakespeare, predict profits, develop video games, and address and communicate about environmental problems or model voter turnout. It all potentially demands quite rigorous STEM thinking, obviating the need for demarcating "proper" applications of STEM in schools. The contributions in this volume point in this direction, implicitly answering Zollman's (2012) call for "STEM literacy for learning," serving as a helpful resource for leaders in STEM education at all levels.

UMass
Amherst, MA, USA Elizabeth H. McEneaney

References

Bybee, R. W. (2013).*The case for STEM education: Challenges and opportunities*. Arlington, VA: NSTA Press.

Ravitch, D. (2003). A brief history of social studies. In J. Leming, L. Ellington, & K. Porter-Magee (Eds.), *Where did social studies go wrong* (pp. 1–5). Washington, DC: Fordham Foundation.

Renyi, J. (2000). Hunting the quark: Interdisciplinary curricula in public schools. In S. Wineburg & P. Grossman (Eds.), *Interdisciplinary curriculum: Challenges to implementation* (pp. 39–56). New York, NY: Teachers College Press.

Shaughnessy, J. M. (2012). STEM: An advocacy position not a content area. *NCTM Summing Up*. February 2.

Weld, J. (2017).*Creating a STEM Culture for Teaching and Learning*. National Science Teachers Association.

Wineburg, S. & Grossman, P. (Eds.) (2000). *Interdisciplinary curriculum: Challenges to implementation.* New York, NY: Teachers College Press.

Zollman, A. (2012). Learning for STEM literacy: STEM literacy for learning. *School Science and Mathematics, 112*(1), 12–19.

Preface

This edited book resulted from our efforts to develop an understanding of the nature of STEM knowledge for our doctoral students and ourselves. It began as a graduate seminar in science education where we explored the natures of the individual STEM disciplines (science, technology, engineering, and mathematics) and research in STEM education alongside our students. The intention was to find overlaps among the characteristics of science, technology, engineering, and mathematics knowledge and develop an idea about the nature of STEM from those overlapping ideas. Over the course of the semester, however, we came to question if there could be a separate nature of STEM knowledge if it is a combination of existing knowledge bases. Further complicating the academic journey was the fact that most STEM research focus on one of the disciplines that comprises STEM itself. We subsequently explored what would STEM teacher education research look like if all the disciplines were truly intertwined and how does this image compare to educators and educational researchers' existing perceptions of STEM. Our journey grew to include teacher educators from different disciplines in higher education institutions across the country. That academic journey was so powerful that we sought to expand the discussion throughout our educational community with this edited book.

This book explores critical questions in STEM education. The questions were prompted by a desire to respond to the educational demands that twenty-first century teachers, and subsequently teacher educators, have had placed on them. When previously they have been teachers of individual disciplines, such as science, math, or technology (and occasionally engineering), they are now often considered STEM teachers. The purpose of the book is to provide a practical resource for teacher educators who seek to prepare teachers to address STEM in a meaningful and interdisciplinary manner. It is not a thorough ontological or epistemological treatment of STEM, although such considerations certainly provide the framework for the writings.

There are three parts within the book, all of which adhere to the definition of STEM as a meaningful interdependence among all disciplines that comprise STEM. In other words, all individual disciplines of STEM are included in ways that are meaningful and showcase the interdependence of the fields. The first part, Nature of the STEM Disciplines, provides the foundation for the discussion of meaningful interdependence by establishing the natures of the component disciplines of STEM (science, technology, engineering, and mathematics). This part does not include epistemological or ontological treatments of the disciplines but rather practical discussion for teaching and research. Concluding this part, the editors explore whether there is a separate STEM discipline with its own nature as well as the challenges and benefits of presuming a nature of STEM. The second part, Critical Questions in Teaching STEM, features applied research on critical questions teacher educators are actively exploring. Chapters in this part showcase their action research, case studies, self-studies, and other classroom-based research connected to learning to effectively prepare classroom teachers to teach STEM in meaningful and interdisciplinary ways. The third part, Critical Questions in STEM, includes chapters that systematically explore and discuss the overall applied constructs of STEM education. These chapters explore such ideas as public perceptions of STEM education, phenomenological case studies on STEM experiences, and content analyses of STEM education documents and texts.

The book you hold is the result of very real and interesting discussions among scholars of teacher education. It includes scholars from all four STEM education disciplines and applied research across these disciplines. Working on this volume has been a very interesting process, and we hope this contribution will be helpful to the fields that comprise STEM and stimulate conversations across the fields.

Bloomington, IN, USA

Valarie L. Akerson
Gayle A. Buck

Contents

About the Editors

Valarie L. Akerson is a Professor of Science Education at Indiana University and a former elementary teacher. Her research focuses on preservice and inservice elementary teachers' ideas about Nature of Science as well as their teaching practices. She is a Past President of the Association for Science Teacher Education and a Past President for NARST: a worldwide organization for improving science teaching and learning through research.

Gayle Buck is an Associate Dean for Research, Development and Innovation as well as a Professor of Science Education. Previously a middle-level science teacher in both urban and rural schools, Professor Buck now teaches courses in science, STEM education, and teacher education. Her research explores (1) student populations traditionally underserved in science education, (2) neglected epistemological assumptions in teaching and learning, and (3) pragmatic and participatory approaches to educational research.

Part I
Nature of the STEM Disciplines

Chapter 1
Nature of Scientific Knowledge and Scientific Inquiry

Norman G. Lederman and Judith Lederman

1.1 Introduction

Before carefully considering how nature of scientific knowledge (NOSK) and scientific inquiry (SI) relate to science, technology, engineering, and mathematics (STEM), it is critical to "define" or explain what is meant by "science." There are many conceptualizations of science. The rotunda in the National Academy of Science contains the following inscription: "To science, pilot of industry, conqueror of disease, multiplier of the harvest, explorer of the universe, revealer of nature's laws, eternal guide to truth. "The quote is not attributed to any individual and the building was built in 1936. It is not clear if the quote is older than 1936. Nobel Prize winning physicist Richard Feynman defined science in the 1970s as "the belief in the ignorance of experts (Feynman & Cashman, 2013). Most recently, Arthur Boucot (famous paleobiologist) in a personal conversation characterized science as "an internally consistent set of lies designed to explain away the universe." These statements are quite varied and as provocative as Boucot's and Feynman's definitions may be they are closer to how science is characterized in recent reform documents, such as the *Next Generation Science Standards* (NGSS Lead States, 2013) and the *National Science Education Standards* (National Research Council, 1996). The question still remains, "what is science?" What conceptualization would be most appropriate for K-12 learners? Commonly, the answer to this question has three parts. First, science is a body of knowledge. This refers to the traditional subjects or body of concepts, laws, and theories. For instance, biology, chemistry, physics etc. The second part refers to how the knowledge is developed. That is scientific inquiry. Inquiry will be discussed in more detail later, but as a student outcome it usually includes the doing of inquiry (e.g., asking questions, developing a design,

N. G. Lederman (✉) · J. Lederman
Illinois Institute of Technology, Chicago, IL, USA
e-mail: ledermann@iit.edu

V. L. Akerson, G. A. Buck (eds.), *Critical Questions in STEM Education*,
Contemporary Trends and Issues in Science Education 51,
https://doi.org/10.1007/978-3-030-57646-2_1

collecting and analyzing data, and drawing conclusions). Additionally, inquiry as a student outcome also includes knowledge about inquiry (e.g., knowing that all investigations begin with a question, there is no single scientific method, research questions guide the procedures, etc.).

Finally, because of the way the knowledge is developed, scientific knowledge has certain characteristics. These characteristics of scientific knowledge are often referred to as nature of scientific knowledge (Lederman, Lederman, & Antink, 2013). Again, these characteristics will be discussed in more detail later, but they usually include, but are not limited to the idea that science is empirically based, involves human creativity, is unavoidably subjective, and is subject to change (Lederman, Wade, & Bell, 1998). Often individuals conflate nature of scientific knowledge (NOSK) with scientific inquiry. Lederman (2007) also notes that the conflation of NOSK and scientific inquiry has plagued research on NOSK from the beginning and, perhaps, could have been avoided by using the phrase "nature of scientific knowledge" as opposed to the more commonly used nature of science (NOS). In this chapter, we will use the term "nature of scientific knowledge" instead of "nature of science" as it more accurately represents its intended meaning (Lederman & Lederman, 2004). Now the critical point is what is the appropriate balance among the three components of science in the science curriculum and science instruction? Current reforms have appropriately recognized that the amount of emphasis has traditionally emphasized the body of knowledge to the detriment of any emphasis on inquiry or nature of scientific knowledge.

Current visions of science education are returning to the perennial goal of scientifically literacy. Again, the roots of scientific literacy and its justification will be discussed in more detail later. But, in general, the goal is to help students use their scientific knowledge to make informed decisions about scientifically based global, societal, or personal decisions. The literate individual can not make such decisions based on scientific knowledge alone. They must also understand the source of the knowledge (i.e., scientific inquiry or the more current term science practices) and the ontological characteristics of the knowledge (i.e., NOSK).

The focus of this chapter is to elaborate on how the interplay among scientific inquiry, NOSK, and STEM may, or may not, contribute to the achievement of scientific literacy. Thus this begs the question of "What is STEM?" For sure STEM has been discussed in each of the chapters in this book. For the sake of brevity, a brief conceptualization follows. STEM has become one of the newest slogans in education, and some critics have noted its ubiquitous and ambiguous use (Bybee, 2013) throughout policy and science education literature. Bybee (2013) coined the phrase "STEM literacy" to make the goal of STEM education more explicit. A STEM approach to science instruction and curriculum incorporates real life problematic situations that require knowledge of nature of scientific knowledge and scientific inquiry, in part, which leads toward the end goal of scientific literacy. Therefore, it could be argued that scientific literacy is the ultimate goal of the integrated STEM approach. It is important to note, here, that contrary to prevalent misconceptions, STEM goes well beyond just placing more emphasis on each of the STEM disciplines. The integration of the STEM disciplines is the intent of the STEM movement. Again, this chapter will focus on whether the interplay of scientific inquiry, nature of scientific knowledge, and STEM can facilitate the development of scientific literacy.

1.2 Scientific Literacy as the Primary Goal of Science Education

Why should our students learn science and to what extent? Are we teaching our students to make them scientists? What happens to those students who do not continue studying science? Don't they need to learn a minimum amount of science? These questions are critical to portray the goal of science education. Science educators believe that the goal of science education is to develop scientific literacy. Since the first use of 'scientific literacy' in the late 1950s, science educators and policy makers have gradually reconceptualized the term to such an extent that one author remarked relatively recently that "scientific literacy is an ill-defined and diffuse concept" (Laugksch, 2000, p. 71). Policy makers and educators often get confused between "science literacy" and "scientific literacy." Often they are considered synonymous, although the two have very different meanings. Science literacy focuses on how much science you know. It is not about applying knowledge and making decisions. "Science literacy" is mostly associated with AAAS Project 2061 (American Association for the Advancement of Science, 1993). In 1985 AAAS, the Carnegie Corporation of New York and the Andrew W. Mellon Foundation launched a project that promised to be radical, ambitious, comprehensive and long-term, in other words, risky and expensive (American Association for the Advancement of Science, 1994). With that philosophy, the program was aptly named "Project 2061." In view of the numerous local, state, and national obstacles and turf infringements, many wondered whether it would take that long to achieve the goals of the program. *Benchmarks for Science Literacy* is the Project 2061 statement of what all students should know and be able to do in science, mathematics, and technology by the end of grades 2, 5, 8, and 12. The recommendations at each grade level suggested reasonable progress toward the adult science literacy goals laid out in the project's 1989 report *Science for All Americans AAAS, 1989). Benchmarks* helped educators decide what to include in (or exclude from) a core curriculum, when to teach it, and why.

On the other hand, "scientific literacy" deals with the aim of helping people use scientific knowledge to make informed decisions. This is a goal that science educators have been striving to achieve, but unfortunately many of us have not truly realized the importance of scientific literacy or might have misrepresented the goal in various platforms. DeBoer (2000) states that the term "scientific literacy" since it was introduced in the late 1950s has defied precise definition. Although it is widely claimed to be a desired outcome of science education, not everyone agrees with what it means.

The goal of science education became formalized at different times in history. After the 1960s the science education community became concerned about the role of science in society, especially given the launching of Sputnik by the Soviet Union in 1957. This event led to a significant increase in funding for science education in an attempt to increase the science pipeline. The primary driving forces were concerns for national security and economic health. In the immediate post-war years, it was proposed that science educators should work to produce citizens who understood science and were sympathetic to the work of scientists (DeBoer, 2000). The U.S. was lacking in producing a workforce who could live and work in such a rapidly changing world. The goals of science teaching, for general education purposes, within this new environment came to be called scientific literacy. According to the Rockefeller Brothers Fund (1958) report, "among the tasks that have increased most frighteningly in complexity is the task of the ordinary citizen who wishes to discharge his civic responsibilities intelligently" (p. 351). The answer was scientific literacy. The Board said:

> Just as we must insist that every scientist be broadly educated, so we must see to it that every educated person be literate in science]…. We cannot afford to have our most highly educated people living in intellectual isolation from one another, without even an elementary understanding of each other's intellectual concern. (p. 369)

The national review of Australian science teaching and learning (Goodrum, Rennie, & Hackling, 2001) defined the attributes of a scientifically literate person. In particular, it stated that a scientifically literate person is (1) interested in and understands the world about him, (2) can identify and investigate questions and draw evidence-based conclusions, (3) is able to engage in discussions of and about science matters, (4) is skeptical and questioning of claims made by others, and (5) can make informed decisions about the environment and their own health and wellbeing.

The current NGSS stresses science practices, but there is very little emphasis on understanding the practices or scientific inquiry and NOSK. Later in this chapter the critical role of scientific inquiry and NOSK for the achievement of scientific literacy will be elaborated in detail. Doing science is necessary as a means, but it should not be the end goal. The end goal should be scientific literacy, which unfortunately is not explicitly mentioned in the standards.

1.3 STEM as a Mechanism to Achieve Scientific Literacy

STEM education must have an educative purpose which goes beyond the slogan "to meet 21st century skills." In the 1990s, the National Science Foundation (NSF) introduced the STEM acronym as an instructional and curricular approach that stresses the integration of science, technology, engineering, and mathematics. But, its ubiquitous and ambiguous use in the education community has created much confusion (Angier, 2010). One of the possible reasons could be the lack of consensus on the meaning of STEM. However, even without a common understanding of

STEM, the development and implementation our STEM curriculum over the years has not been deterred. Bybee (2013) addressed four components of STEM literacy. STEM literacy refers to an individual's

- knowledge, attitudes, and skills to identify questions and problems in life situations, explain the natural and designed world, and draw evidence-based conclusions about STEM related-issues
- understanding of the characteristic features of STEM disciplines as forms of human knowledge, inquiry, and design;
- awareness of how STEM disciplines shape our material, intellectual, and cultural environments; and
- willingness to engage in STEM-related issues and with the ideas of science, technology, engineering, and mathematics as a constructive, concerned, and reflective citizen.

From the above components of STEM literacy, it is evident that students need to have experiences to apply their knowledge and skills. But the debate over other aspects of STEM education has not been settled yet. For instance, is STEM a separate discipline or just an integrated curriculum approach? The idea of considering STEM as a separate discipline has been a puzzle for many science educators. STEM disciplines are all different ways of knowing and have different conventions for what constitutes data and evidence. STEM is an integrated curriculum approach, but because it deals with different ways of knowing, true integration is never achieved; just an interdisciplinary connection. Individual STEM disciplines "are based on different epistemological assumptions" and integration of the STEM subjects may detract from the integrity of any individual STEM subject (Williams, 2011, p. 30). If STEM is conceptualized as a curriculum approach, its interdisciplinary nature entails not just the acquisition and application of scientific knowledge, but also the other knowledge bases. Wang, Moore, Roehrig, and Park (2011) explained that *interdisciplinary integration* begins with a real-world problem. It incorporates cross-curricular content with critical thinking, problem-solving skills, and knowledge in order to reach a conclusion. Students engage themselves in different real-life STEM related personal and societal situations to make informed decisions. More specifically, STEM curriculum in classrooms and programs can ensure five skill sets including adaptability, complex communications, nonroutine problem solving, self-management, and systems thinking (NRC, 2008). The National Research Council (2010) elaborated on these five skills in its report, *Exploring the Intersection of Science Education and 21st-Century Skills*. Furthermore, in a second report (NRC, 2012), *Education for Life and Work: Developing Transferable Knowledge and Skills in the 21st Century* it was emphasized that these 21st century skills are necessary if students are to solve the personal and societal problems. This is what it means to be an informed citizen. If we put the components of scientific literacy alongside STEM in terms of science instruction, it can be argued that both focus on the context of the world we live in and the decisions we make in everyday life. Those decisions are not just based on science. Different social, political, cultural perspectives are all part of these decisions. While making those decisions, people are supposed to apply some of their other knowledge bases such as mathematical reasoning and technological and engineering processes. For example, if individuals are supposed to make any decisions about whether wind or solar energy is best for the environment and economy, it must be kept in mind that the solution is not just based on scientific knowledge, but also knowledge of other technical or engineering features that explain how these two types of energy sources actually operate. Further, mathematical knowledge is needed to be able to calculate the economic efficiency of the two sources of energy. Can we imagine any activity that requires this type of decision making as a part of the STEM curricular approach? The answer is clearly yes. Thus, it can be argued that STEM as an instructional and curricular approach is consistent with the idea of scientific literacy.

1.4 The Role of Scientific Inquiry in Science Education

As previously discussed, the unclear definitions and multiple uses of the phrase "scientific literacy" resulted in much confusion. However, the phrase "scientific inquiry" is guilty of the same. What it means has been elusive and it is at least one of the reasons why the *Next Generation Science Standards* (NGSS Lead States, 2013) emphasizes "science practices" as opposed to scientific inquiry. The *National Science Education Standards ([NSES]* National Research Council, 1996*)* arguably made the most concerted effort to unpack the meaning of scientific inquiry. The NSES envisioned scientific inquiry as both subject matter and pedagogy in its three part definition. However, with all the effort, confusion remained and the National Research Council had to develop an addendum of sorts, a few years later, titled *Inquiry and the National Science Education Standards* (NRC, 2000). On the one hand, scientific inquiry was conceptualized as a teaching approach. That is, the science teacher would engage students in situations (mostly open-ended) they could ask questions, collect data, and draw conclusions. In short, the purpose of the teaching approach was to enable students to learn science subject matter in a manner similar to how scientists do their work. Although closely related to science processes, scientific inquiry extends beyond the mere development of process skills such as observing, inferring, classifying, predicting, measuring, questioning, interpreting and analyzing data. Scientific inquiry includes the traditional science processes, but also refers to the combining of these processes with scientific knowledge, scientific reasoning and critical thinking to develop scientific knowledge. From the perspective of the *National Science Education Standards* (NRC, 1996), students are expected to be able to develop scientific questions and then design and conduct investigations that will yield the data necessary for arriving at answers for the stated questions.

Scientific inquiry, in short, refers to the systematic approaches used by scientists in an effort to answer their questions of interest. Pre-college students, and the general public for that matter, believe in a distorted view of scientific inquiry that has resulted from schooling, the media, and the format of most scientific reports. This distorted view is called THE SCIENTIFIC METHOD. That is, a fixed set and sequence of steps that all scientists follow when attempting to answer scientific questions. A more critical description would characterize THE METHOD as an algorithm that students are expected to memorize, recite, and follow as a recipe for success. The visions of reform, as well as any study of how science is done, are quick to indicate that there is no single fixed set or sequence of steps that all scientific investigations follow. The contemporary view of scientific inquiry advocated is that the research questions guide the approach and the approaches vary widely within and across scientific disciplines and fields (Lederman et al., 1998).

The perception that a single scientific method exists owes much to the status of classical experimental design. Experimental designs very often conform to what is presented as THE SCIENTIFIC METHOD and the examples of scientific investigations presented in science textbooks most often are experimental in nature. The problem, of course, is not that investigations consistent with "the scientific method" do not exist. The problem is that experimental research is not representative of scientific investigations as a whole. Consequently, a very narrow and distorted view of scientific inquiry is promoted in our K-12 science curriculum.

At a general level, scientific inquiry can be seen to take several forms (i.e., descriptive, correlational, and experimental). Descriptive research is the form of research that often characterizes the beginning of a line of research. This is the type of research that derives the variables and factors important to a particular situation of interest. Whether descriptive research gives rise to correlational approaches depends upon the field and topic. For example, much of the research in anatomy and taxonomy are descriptive in nature and do not progress to experimental or correlational types of research. The purpose of research in these areas is very often simply to describe. On the other hand, there are numerous examples in the history of anatomical research that have lead to more than a description. The initial research concerning the cardiovascular system by William Harvey was descriptive in nature. However, once the anatomy of blood vessels had been described, questions arose concerning the circulation of blood through the vessels. Such questions lead to research that correlated anatomical structures with blood flow and experiments based on models of the cardiovascular system (Lederman et al., 1998).

To briefly distinguish correlational from experimental research, the former explicates relationships among variables identified in descriptive research and experimental research involves a planned intervention and manipulation of the variables studied in correlational research in an attempt to derive causal relationships. In some cases, lines of research can been seen to progress from descriptive to correlational to experimental, while in other cases (e.g., descriptive astronomy) such a progression is not necessarily possible. This is not to suggest, however, that the experimental design is more scientific than descriptive or correlational designs but instead to clarify that there is not a single method applicable to every scientific question.

Scientific inquiry has always been ambiguous in its presentation within science education reforms. In particular, inquiry is perceived in three different ways. It can be viewed as a set of skills to be learned by students and combined in the performance of a scientific investigation. It can also be viewed as a cognitive outcome that students are to achieve. In particular, the current visions of reform (e.g. NGSS Lead States, 2013; NRC, 1996) are very clear (at least in written words) in distinguishing between the performance of inquiry (i.e., what students will be able to do) and what students know about inquiry (i.e., what students should know). For example, it is one thing to have students set up a control group for an experiment, while it is another to expect students to understand the logical necessity for a control within an experimental design. Unfortunately, the subtle difference in wording noted in the reforms (i.e., "know" versus "do") is often missed by everyone except the most careful reader. The third use of "inquiry" in reform documents relates strictly to pedagogy and further muddies the water. In particular, current wisdom advocates that students learn science best through an inquiry-oriented teaching approach. It is believed that students will best learn scientific concepts by doing science (NGSS Lead States, 2013).

In this sense, "scientific inquiry" is viewed as a teaching approach used to communicate scientific knowledge to students (or allow students to construct their own knowledge) as opposed to an educational outcome that students are expected to learn about and learn how to do. Indeed, it is the pedagogical conception of inquiry that it is unwittingly communicated to most teachers by science education reform documents, with the two former conceptions lost in the shuffle. Although the processes that scientists use when doing inquiry (e.g. observing, inferring, analyzing data, etc.) are readily familiar to most, knowledge *about* inquiry, as an instructional outcome is not. This is the perspective of inquiry that distinguishes current reforms from those that have previously existed, and it is the perspective on inquiry that is not typically assessed. In summary, the knowledge *about* inquiry included in current science education reform efforts includes the following (NGSS Lead States, 2013, NRC, 1996):

- Scientific investigations all begin with a question, but do not necessarily test a hypothesis
- There is no single set and sequence of steps followed in all scientific investigations (i.e., there is no single scientific method)
- Inquiry procedures are guided by the question asked
- All scientists performing the same procedures may not get the same results
- Inquiry procedures can influence the results
- Research conclusions must be consistent with the data collected
- Scientific data are not the same as scientific evidence
- Explanations are developed from a combination of collected data and what is already known

1.5 Scientific Inquiry as a Component of Scientific Literacy and Its Relationship to STEM

Although scientific inquiry has been viewed as an important educational outcome for science students for over 100 years, it was Showalter's (1974) work that galvanized scientific inquiry, as well as NOSK, important components within the over arching framework of scientific literacy. As previously discussed, the phrase scientific literacy had been discussed by numerous authors before Showalter (Dewey, 1916; Hurd, 1958; National Education Association, 1918, 1920; National Society for the Study of Education, 1960, among others), it was his work that clearly delineated the dimensions of scientific literacy in a manner that could easily be translated into objectives for science curricula. Showalter's framework consisted of the following seven components:

- **Nature of Science** – The scientifically literate person understands the nature of scientific knowledge.
- **Concepts in Science** – The scientifically literate person accurately applies appropriate science concepts, principles, laws, and theories in interacting with his universe.
- **Processes of Science** – The scientifically literate person uses processes of science in solving problems, making decisions and furthering his own understanding of the universe.
- **Values** – The scientifically literate person interacts with the various aspects of how universe in a way that is consistent with the values that underlie science.
- **Science-Society** – The scientifically literate person understands and appreciates the joint enterprise of science and technology and the interrelationships of these with each other and with other aspects of society.
- **Interest** – The scientifically literate person has developed a richer, more satisfying, and more exciting view of the universe as a result of his science education and continues to extend this education throughout his life.
- **Skills** – The scientifically literate person has developed numerous manipulative skills associated with science and technology.

(Showalter, 1974, p. 1–6)

Science processes (now known as inquiry or practices), and NOSK) were clearly emphasized. The attributes of a scientifically literate individual were later reiterated by the National Science Teachers Association [NSTA] (1982). The NSTA dimensions of scientific literacy were a bit expanded from Showalter's and included:

- Uses science concepts, process skills, and values making responsibly everyday decisions;
- Understands how society influences science and technology as well as how science and technology influence society;
- Understands that society controls science and technology through the allocation of resources;

- Recognizes the limitations as well as the usefulness of science and technology in advancing human welfare;
- Knows the major concepts, hypotheses, and theories of science and is able to use them;
- Appreciates science and technology for the intellectual stimulus they provide;
- Understands that the generation of scientific knowledge depends on inquiry process and conceptual theories;
- Distinguishes between scientific evidence and personal opinion;
- Recognizes the origin of science and understands that scientific knowledge is tentative, and subject to change as evidence accumulates;
- Understands the application of technology and the decisions entailed in the use of technology;
- Has sufficient knowledge and experience to appreciate the worthiness of research and technological developments;
- Has a richer and more exciting view of the world as a result of science education; and
- Knows reliable sources of scientific and technological information and uses these sources in the process of decision making.

The importance of scientific inquiry, or practices as it is called in the NGSS, as a critical component of scientific literacy should be clear.

STEM, in current conceptions, is characterized as an integrated approach to curriculum that addresses the interactions of science, technology, engineering, and mathematics to solve problems in a more authentic manner than the current curriculum approach. That is, the typical science curriculum has perennially separated the various disciplines during precollege instruction, not to mention the exclusion of any formal attention to technology or engineering. Current questions about the natural world and/or societal or personal issues are more commonly not the purview of any singular discipline, but rather require the collaboration of various individuals, working in a team, with various backgrounds and expertise. This is the nature of STEM. We are not saying that STEM is a discipline with its own "nature" as in nature of science. We are merely characterizing STEM as a curriculum approach.

1.6 Understanding Nature of Scientific Knowledge as a Goal of Science Education and Its Relationship to Scientific Literacy

The relationship and differences between nature of scientific knowledge (NOSK) and nature of scientific inquiry (SI) is often discussed and confused within existing literature (Lederman & Lederman, 2014). NOSK, as opposed to the more popular nature of science (NOS) is used here to be more consistent with the original meaning of the construct (Lederman, 2007).

Given the manner in which scientists develop scientific knowledge (i.e., SI), the knowledge is engendered with certain characteristics. These characteristics are what typically constitute NOS (Lederman, 2007). As mentioned before there is a lack of consensus among scientists, historians of science, philosophers of science, and science educators about the particular aspects of NOSK. This lack of consensus, however, should neither be disconcerting nor surprising given the multifaceted nature and complexity of the scientific endeavor. Conceptions of NOS have changed throughout the development of science and systematic thinking about science and are reflected in the ways the scientific and science education communities have defined the phrase "nature of science" during the past 100 years (e.g., AAAS, 1990, 1993; Central Association for Science and Mathematics Teachers, 1907; Klopfer & Watson, 1957; NSTA, 1982).

However, many of the disagreements about the definition or meaning of NOSK that continue to exist among philosophers, historians, and science educators are irrelevant to K-12 instruction. The issue of the existence of an objective reality as compared to phenomenal realities is a case in point. There is an acceptable level of generality regarding NOS that is accessible to K-12 students and relevant to their daily lives. Moreover, at this level, little disagreement exists among philosophers, historians, and science educators. Among the characteristics of the scientific enterprise corresponding to this level of generality are that scientific knowledge is tentative (subject to change), empirically-based (based on and/or derived from observations of the natural world), subjective (theory-laden), necessarily involves human inference, imagination, and creativity (involves the invention of explanations), and is socially and culturally embedded. Two additional important aspects are the distinction between observations and inferences, and the functions of, and relationships between scientific theories and laws. What follows is a brief consideration of these characteristics of science and scientific knowledge.

First, students should be aware of the crucial distinction between observation and inference. Observations are descriptive statements about natural phenomena that are "directly" accessible to the senses (or extensions of the senses) and about which several observers can reach consensus with relative ease. For example, objects released above ground level tend to fall and hit the ground. By contrast, inferences are statements about phenomena that are not "directly" accessible to the senses. For example, objects tend to fall to the ground because of "gravity." The notion of gravity is inferential in the sense that it can *only* be accessed and/or measured through its manifestations or effects. Examples of such effects include the perturbations in predicted planetary orbits due to inter-planetary "attractions," and the bending of light coming from the stars as its rays pass through the sun's "gravitational" field.

Second, closely related to the distinction between observations and inferences is the distinction between scientific laws and theories. Individuals often hold a simplistic, hierarchical view of the relationship between theories and laws whereby theories become laws depending on the availability of supporting evidence. It follows from this notion that scientific laws have a higher status than scientific theories. Both notions, however, are inappropriate because, among other things, theories and laws are different kinds of knowledge and one can not develop or be transformed into the other. Laws are *statements or descriptions of the relationships* among observable phenomena. Boyle's law, which relates the pressure of a gas to its volume at a constant temperature, is a case in point (Lederman et al., 1998).

Theories, by contrast, *are inferred explanations* for observable phenomena. The kinetic molecular theory, which explains Boyle's law, is one example. Moreover, theories are as legitimate a product of science as laws. Scientists do not usually formulate theories in the hope that one day they will acquire the status of "law." Scientific theories, in their own right, serve important roles, such as guiding investigations and generating new research problems in addition to explaining relatively huge sets of seemingly unrelated observations in more than one field of investigation. For example, the kinetic molecular theory serves to explain phenomena that relate to changes in the physical states of matter, others that relate to the rates of chemical reactions, and still other phenomena that relate to heat and its transfer, to mention just a few.

Third, even though scientific knowledge is, at least partially, based on and/or derived from observations of the natural world (i.e., empirical), it nevertheless involves human imagination and creativity. Science, contrary to common belief, is not a totally lifeless, rational, and orderly activity. Science involves the *invention* of explanations and this requires a great deal of creativity by scientists. The "leap" from atomic spectral lines to Bohr's model of the atom with its elaborate orbits and energy levels is a case in point. This aspect of science, coupled with its inferential nature, entails that scientific concepts, such as atoms, black holes, and species, are functional theoretical models rather than faithful copies of reality.

Fourth, scientific knowledge is subjective or theory-laden. Scientists' theoretical commitments, beliefs, previous knowledge, training, experiences, and expectations actually influence their work. All these background factors form a *mind-set* that *affects* the problems scientists investigate and how they conduct their investigations, what they observe (and do not observe), and how they make sense of, or interpret their observations. It is this (sometimes collective) individuality or mind-set that accounts for the role of subjectivity in the production of scientific knowledge. It is noteworthy that, contrary to common belief, science never starts with neutral observations (Chalmers, 1982). Observations (and investigations) are always motivated and guided by, and acquire meaning in reference to questions or problems. These questions or problems, in turn, are derived from within certain theoretical perspectives.

Fifth, science as a human enterprise is practiced in the context of a larger culture and its practitioners (scientists) are the product of that culture. Science, it follows, affects and is affected by the various elements and intellectual spheres of the culture in which it is embedded. These elements include, but are not limited to, social fabric, power structures, politics, socioeconomic factors, philosophy, and religion. An example may help to illustrate how social and cultural factors impact scientific knowledge. Telling the story of the evolution of humans (*Homo sapiens*) over the course of the past seven million years is central to the biosocial sciences. Scientists have formulated several elaborate and differing story lines about this evolution.

Until recently, the dominant story was centered about "the man-hunter" and *his* crucial role in the evolution of humans to the form we now know (Lovejoy, 1981).

This scenario was consistent with the white-male culture that dominated scientific circles up to the 1960s and early 1970s. As the feminist movement grew stronger and women were able to claim recognition in the various scientific disciplines, the story about hominid evolution started to change. One story that is more consistent with a feminist approach is centered about "the female-gatherer" and *her* central role in the evolution of humans (Hrdy, 1986). It is noteworthy that both story lines are consistent with the available evidence.

Sixth, it follows from the previous discussions that scientific knowledge is never absolute or certain. This knowledge, including "facts," theories, and laws, is tentative and subject to change. Scientific claims change as new evidence, made possible through advances in *theory* and technology, is brought to bear on existing theories or laws, or as old evidence is reinterpreted in the light of new theoretical advances or shifts in the directions of established research programs. It should be emphasized that tentativeness in science does not only arise from the fact that scientific knowledge is inferential, creative, and socially and culturally embedded. There are also compelling logical arguments that lend credence to the notion of tentativeness in science. Indeed, contrary to common belief, scientific hypotheses, theories, and laws can *never* be absolutely "proven." This holds irrespective of the amount of empirical evidence gathered in the support of one of these ideas or the other (Popper, 1963, 1988). For example, to be "proven," a certain scientific law should account for *every single instance* of the phenomenon it purports to describe *at all times*. It can logically be argued that one such future instance, of which we have no knowledge whatsoever, may behave in a manner contrary to what the law states. As such, the law can never acquire an absolutely "proven" status. This equally holds in the case of hypotheses and theories.

It is clear from the attributes of a scientifically literate individual espoused by Showalter (1974) and NSTA (1982), that NOSK is considered a critical component of scientific literacy. If precollege and postsecondary students are expected to make informed decisions about scientifically based personal and societal issues they must have an understanding of the sources and limits of scientific knowledge. For example, it is becoming increasingly common for the public to hear alternative viewpoints presented by scientists on the same topic. Are organic foods healthier to eat? Should GMOs be avoided at all costs or are they perfectly safe? Is drinking water with a pH of approximately 7.3 healthier than drinking water that is more alkaline or more acidic? In Asia it is believed that the ingestion of cold liquids puts a stress on your body and should be avoided. Consequently, it is not uncommon to find drinking fountains that provide warm and hot water as opposed to the cold water provided by drinking fountains in most regions throughout the world. You can find qualified scientists arguing both sides of the aforementioned issues. Sometimes the claims are based on pseudoscience, like current claims that there really is no global warming or the claim that biological evolution never occurred. Alternatively, these differences in perspectives and knowledge are the result of science in action. It is the results of the nature of scientific knowledge. Science is done by humans and it is limited, or strengthened by the foibles that all humans have. Scientific knowledge is tentative, or subject to change. We never have all of the data, and if we did we would not know it. If you look up in the sky on a clear night you will see a white, circular object. We would all agree that the object is the moon. Three hundred years ago if we looked at the same object we would call it a planet. This is because the current view of our solar system is guided by heliocentric theory. This theory places the sun at the center of the solar system and any objects orbiting the sun is a planet (e.g., the earth) and any object orbiting a planet is a moon or satellite. Three hundred years ago our view was guided by the geocentric theory which places the earth at the center and anything orbiting the earth was considered a planet (e.g., our current moon). The objects and observations have not changed, but our interpretation has because of a change in the theories we adopt. You could say that our theories "bias" our interpretations of data. Scientists make observations, but then eventually make inferences because all the data are not accessible through our senses. This is why scientific knowledge is tentative and partly a function of human subjectivity and creativity. The examples illustrating the characteristics of scientific knowledge (i.e., NOSK) are endless and an understanding of these characteristics is critical when making decisions on scientifically based issues.

1.7 The Promise of STEM and the Achievement of Scientific Literacy

Given the previous discussions about inquiry, NOSK, STEM, and scientific literacy, it seems quite logical to assume that revising our curricular approach to be more consistent with STEM, and the vision of the NGSS, would enhance our ability to enhance the scientific literacy of our precollege and postsecondary students. After all, a STEM approach seems to be a more authentic because it does not pigeonhole the issues our citizens face into discrete discipline "silos." Indeed, none of the really significant issues that affect us as a global community, society, culture, or individually are the purview of any single discipline. Further, it can be argued that none of the significant scientifically based issues we face are limited to the STEM fields. Isn't this why we see additionally permutations of STEM, such as STEAM? In summary, STEM provides the scientific and technical knowledge, while scientific inquiry and NOSK provides us with knowledge about how the subject matter is developed (inquiry) and the unavoidable characteristics (NOSK) derived from how the knowledge was developed.

Logic is one thing, but what do we know and what do we need to know? Is there strong empirical support to show that students exposed to STEM exhibit increased achievement, critical thinking, and problem solving ability? It seems the first place to look is at the research on integrated instruction (see Czerniak, 2007; Czerniak & Johnson, 2014). The idea of integration has existed for over 100 years, and it mainly focused on the integration of science and mathematics. In the past decade there has been an increase in empirical research mainly because of the emergence of STEM and the NGSS. In general, empirical support for integrated instruction is mixed at best. It is important to note that "integration" has many different meanings and that none of the research systematically has focused on the integration of science and engineering, although engineering projects have often been included in traditional science courses.

There are definite obstacles to using STEM to achieve scientific literacy. Some are general, but others are specific to NOSK and scientific inquiry. At the general level is the issue of teacher preparation. The current approach to the education of teachers is specific to the particular disciplinary licensure. That is, teachers are prepared to become biology, mathematics, physics, chemistry, and earth science teachers, among others. Given the volumes of research on pedagogical content knowledge and discipline specific pedagogy using a "generalist" is not advisable. Consequently, either licensure programs will need to be changed or STEM will have to integrate learning through team teaching or a middle school model. In either case, the obstacles are huge.

Will there be a capstone STEM course or will STEM be included in every course? If it is included in every course, then obviously the "home" discipline (e.g., chemistry) will be emphasized over the other STEM disciplines. This is hardly true integration. If STEM is seen as a capstone course, the road forward will be easier, although the licensure of teachers previously discussed remains a problem.

Let us not forget that the focus of this chapter is on scientific inquiry and NOSK. And it is in this area that STEM is most problematic. Theoretically, the rationale for the STEM approach is to enable students to more authentically engage in real world problems of interest that enable them to learn the subject matter of the STEM disciplines and demonstrate the decision-making skills evident in a scientifically literate individual. Such a curriculum or instructional approach most obviously focuses on problem solving and critical thinking. Whenever disciplines are integrated the nature of the disciplines and how disciplinary knowledge is developed. This brings us back to inquiry and NOSK. For example, science attempts to answer questions about the natural world. It does not try to produce any a priori outcomes. Engineering, on the other hand, attempts to produce certain effects. Surely, engineering and the sciences are closely related, but they are different. Science never claims to arrive at absolute truths. All knowledge is subject to change. However, mathematics can arrive at absolute proofs in the mathematical world that it has created. Science must test its knowledge against the natural world, it is empirically based. Mathematics is not necessarily empirically based, it has imaginary numbers. When dealing with lower level knowledge you can integrate the knowledge around relevant themes. However, when it comes to higher level applications and decisions, the conventions of inquiry, the conventions of what constitutes evidence, and the ontological status of the knowledge differ. In true integration disciplinary knowledge of one way of knowing is not privileged above another. It seems with such different ways of knowing, the obstacle of STEM may be insurmountable when it comes to issues of inquiry and NOSK.

1.8 A Needed Research Agenda

It is clear that we know very little about the promise of STEM enhancing scientific literacy. There is much research that needs to be done with respect to all aspects of STEM. Specifically, with respect to NOSK and scientific inquiry, the following needs to be investigated:

- Can effective models of teacher education be developed that enable teachers to simultaneously honor significantly different ways of knowing in a single course?
- When students are designing an investigation how do they negotiate the differing conventions of data collection and interpretation across the STEM fields?
- How are differing conclusions for an investigation handled? Are they characterized as unavoidable differences in interpretation and research design or is it concluded that there is only one solution?
- As students work in groups during an investigation or project, on what basis are decisions made when differences in opinion arise?
- As students are expected to learn NOSK, nature of engineering, nature of mathematics, and nature of technology, does knowledge of one of these impact, negatively or positively, analogous knowledge in another field?

References

American Association for the Advancement of Science. (1990). *Science for all Americans*. New York: Oxford University Press.

American Association for the Advancement of Science. (1993). *Benchmarks for science literacy*. New York: Oxford University Press.

American Association for the Advancement of Science. (1994). *Benchmarks for science literacy*. New York: Oxford University Press.

Angier, N. (2010). STEM education has little to do with flowers. *The New York Times*, D2.

Bybee, R. W. (2013). *The case for STEM education: Challenges and opportunities*. Washington, DC: NSTA Press.

Central Association for Science and Mathematics Teachers. (1907). A consideration of the principles that should determine the courses in biology in secondary schools. *School Science and Mathematics, 9*(3), 241–247.

Chalmers, A. F. (1982). *What is this thing called science?* (2nd Edn.). Queensland, Australia: University of Queensland Press.

Czerniak, C. M. (2007). Interdisciplinary science teaching. In S. K. Abell & N. G. Lederman (Eds.), *Handbook of research on science education* (pp. 537–560). Mahwah, NJ: Lawrence Erlbaum Publishing.

Czerniak, C. M., & Johnson, C. C. (2014). Interdisciplinary science teaching. In I. N. G. Lederman & S. K. Abell (Eds.), *Handbook of research on science education* (pp. 395–411). New York: Routledge.

DeBoer, G. E. (2000). Scientific literacy: Another look at its historical and contemporary meanings and its relationship to science education reform. *Journal of Research in Science Teaching: The Official Journal of the National Association for Research in Science Teaching, 37*(6), 582–601.

Dewey, J. (1916). *Democracy and education: An introduction to the philosophy of education*. New York: MacMillan.

Feynman, R. P., & Cashman, D. (2013). *The pleasure of finding things out*. Blackstone Audio, Incorporated.

Fund, R. B. (1958). *The pursuit of excellence: Education and the future of America; panel report V of the special studies project*. Doubleday.

Goodrum, D., Rennie, L. J., & Hackling, M. W. (2001). *The status and quality of teaching and learning of science in Australian schools: A research report*. Canberra, Australia: Department of Education, Training and Youth Affairs.

Hrdy, S. B. (1986). Empathy, polyandry, and the myth of the coy female. In Bleier, R. (Ed.), *Feminist approaches to science* (pp. 119–146). New York: Pergamon.

Hurd, P. D. (1958). Science literacy: Its meaning for American schools. *Educational Leadership, 16*, 13–16.

Klopfer, L. E., & Watson, F. G. (1957). Historical materials and high school science teaching. *The Science Teacher, 24*(6), 264–293.

Laugksch, R. C. (2000). Scientific literacy: A conceptual overview. *Science Education, 84*(1), 71–94.

Lederman, N. G., Wade, P., & Bell, R. L. (1998). Assessing understanding of the nature of science: A historical perspective. In W. McComas (Ed.), *The nature of science in science education* (pp. 331–350). Dordrecht, the Netherlands: Springer.

Lederman, N. G., & Lederman, J. S. (2004). The nature of science and scientific inquiry. *The art of teaching science*, 2–17.

Lederman, N. G., & Lederman, J. S. (2014). Research on teaching and learning of nature of science. In N. G. Lederman, & S. K. Abell (Eds.), *Handbook of research on science education* (Vol. II, pp. 600–620). New York: Routledge.

Lederman, N. G. (2007). Nature of science: Past, present, and future. In N. G. Lederman & S. K. Abell (Eds.), *Handbook of research on science education* (pp. 831–879). Mahwah, NJ: Lawrence Erlbaum Publishing.

Lederman, N. G., Lederman, J. S., & Antink, A. (2013). Nature of science and scientific inquiry as contexts for the learning of science and achievement of scientific literacy. *International Journal of Education in Mathematics, Science and Technology, 1*(3), 138.

Lovejoy, C. O. (1981). The origin of man. *Science, 211*, 341–350.

National Education Association. (1918). *Cardinal principles of secondary education: A report of the commission on the reorganization of secondary education*. (U.S. Bureau of Education Bulletin No. 35). Washington, DC: U.S. Government Printing Office.

National Education Association. (1920). *Reorganization of science in secondary schools: A report of the commission on the reorganization of secondary education*. (U.S. Bureau of Education Bulletin No. 20). Washington, DC: U.S. Government Printing Office.

National Research Council. (1996). *National science education standards*. Washington, DC: National Academy Press.

National Research Council. (2000). *Inquiry and the national science education standards*. Washington, DC: National Academy Press.

National Research Council. (2008). *Research on future skill demands: A workshop summary*. Washington, DC: National Academies Press.

National Research Council. (2010). *Exploring the intersection of science education and 21st century skills: A workshop summary*. Washington, DC: National Academies Press.

National Research Council. (2012). *A framework for K–12 science education: Practices, crosscutting concepts, and core ideas*. Washington, DC: National Academies Press.

National Science Teachers Association. (1982). *Science-technology-society: Science education for the 1980s*. Washington, DC: Author.

National Society for the Study of Education. (1960). *Rethinking science education: Yearbook of the national society for the study of education* (Vol. 59, p. 113). Chicago: University of Chicago Press.

NGSS Lead States. (2013). *Next generation science standards: For states, by states*. Washington, DC: National Academies Press.

Popper, K. R. (1963). *Conjectures and refutations: The growth of scientific knowledge*. London: Routledge.
Popper, K. R. (1988). *The open universe: An argument for indeterminism*. London: Routledge.
Showalter, V. M. (1974). What is unified science education? Program objectives and scientific literacy. *Prism, 2*(3–4), 1–6.
Wang, H., Moore, T., Roehrig, G., & Park, M. (2011). STEM integration: Teacher perceptions and practice. *Journal of Pre-College Engineering Education Research, 1*(2), 1–13.
Williams, J. (2011). STEM education: Proceed with caution. *Design and Technology Education, 16*(1), 26–35.

Norman G. Lederman is Chair and Distinguished Professor of Mathematics and Science Education at the Illinois Institute of Technology. Dr. Lederman received his Ph.D. in Science Education and he possesses MS degrees in both Biology and Secondary Education. Prior to his 35 + years in science teacher education, Dr. Lederman was a high school teacher of biology and chemistry for 10 years. Dr. Lederman is internationally known for his research and scholarship on the development of students' and teachers' conceptions of nature of science and scientific inquiry. He has been author or editor of 12 books, written 25 book chapters, published over 220 articles in professional journals, and made over 500 presentations at professional conferences around the world. He is the Co-Editor of Volume I, II, and III (forthcoming) of the Handbook of Research on Science Education and was the previous co-editor of the *Journal of Science Teacher Education* and *School Science and Mathematics*. Dr. Lederman is a former President of the National Association for Research in Science Teaching (NARST) and the Association for the Education of Teachers in Science (AETS, now known as ASTE). He has also served as Director of Teacher Education for the National Science Teachers Association (NSTA). He has been named a Fellow of the American Association for the Advancement of Science and the American Educational Research Association, and has received the Distinguished Contributions to Science Education through Research Award from the National Association for Research in Science Teaching.

Judith S. Lederman is an Associate Professor and Director Of Teacher Education at Illinois Institute of Technology and a former museum curator and secondary physics and biology teacher. Her research focuses on the teaching, learning and assessing of Nature of Science and Scientific Inquiry in formal and informal settings. She has served on the boards of NARST, NSTA and ASTE.

Chapter 2
The Nature of Technology

Theresa A. Cullen and Meize Guo

2.1 Introduction

Technology is both the tools that are used but also the systematic processes by which problems are solved. For example, in biology, technology is used to coordinate efforts to find vaccines by allowing for meaningful communication and data sharing. Meanwhile, in engineering, technology allows calculations that could not be done before in order to design structures and solutions. In math, technology serves to speed up processing of calculations to allow for greater complexity and to provide application and visualization of mathematical models.

The nature of technology is important as we develop our way of knowing in our ever increasingly technological society and use its affordances to solve problems and create solutions in science, engineering and math. Thinking about technology is difficult since people often view it through different perspectives. Like the nature of science, technology views are shaped by individuals' experiences and cultures, thus affecting their view. They may be focused on the practical uses of technology versus focusing theoretically on technology's role in our lives. To discuss the nature of technology, we must first examine how technology is defined and then discuss how it is applied in educational settings. Through this examination we can develop a deeper meaning of technology as it relates to learning and integration across the STEM fields. Technology is an integral part of the STEM acronym because it provides the tools and processes by which the other areas advance and do their work.

T. A. Cullen (✉)
Jeannine Rainbolt College of Education, University of Oklahoma, Norman, OK, USA

Arkansas Tech University, Russellville, Arkansas, USA
e-mail: tcullen@atu.edu

M. Guo
Indiana University, Bloomington, IN, USA

V. L. Akerson, G. A. Buck (eds.), *Critical Questions in STEM Education*,
Contemporary Trends and Issues in Science Education 51,
https://doi.org/10.1007/978-3-030-57646-2_2

Processes that are developed in one STEM field are often influenced by and shared via technology. For example, models used in biology to for population evolution rely on computational modeling developed within the math field and influence models used in engineering to design solutions and environments. A key to meaningfully integrating technology across STEM is to look for places where technology and other STEM fields share a way of knowing. By doing that, we can integrate not only the technology tools in our teaching and learning of STEM topics but also engage in explicit reflection on the role of technology in our lives and communities. Moreover, the interdependence between technology and science, engineering, mathematics shall emerge in this process as well.

2.2 Definitions of Technology

Technology is often understood to mean different things to different people. Not surprisingly, multiple, nuanced definitions of technology exist that reflect its historical development. For some technology is the processes used to solve problems and for others it is the tools like computers and calculators that allow for complex calculations and modeling in other STEM fields. Scholars and technology researchers have struggled to define technology for a long time but lack consensus. Major ideas center around process, social impact, and how computers interact with humans.

Zvorikine (1961) summarized that technology was defined to indicate art and craftsmanship, but also meant procedure, methods, formulas, and with the development of machinery production, it primarily referred to the labor. Beyond the concepts of technology as artifacts and technology as procedure, Pacey (1983) emphasized the social impacts of technology. Technology develops and interacts with multiple dimensions of society at the individual, social, political, cultural and economic levels. Similarly, Waight and Abd-El-Khalick (2012) state the nature of technology should include an individual's culture and values as well. This has important implications as technology has evolved. For example, Mitcham (1994) claimed the intention of user can help identify and shape technology from the human computer interaction (HCI) perspective. In HCI, technology is ubiquitous and often wearable, which creates an interesting lens with which to view the social impacts of technology in the modern world. Given new technologies such as the Internet of Things (IOT) where smart technology helps to predict our needs and report our use back to companies. Examining the social impact of technology is growing to be a more important philosophical and practical question (Berman & Cerf, 2017). Technologies like IOT depend on mathematics to model interactions and engineering to apply these technologies to new designs to improve peoples' lives.

Technology scholars acknowledge the challenge in defining the term based on these different lenses through which technology is viewed. Tiles and Oberdiek (2013) identified three difficulties when defining "technology." The first difficulty is that any proposed definition of technology cannot exclude other definitions. Because of different backgrounds and perceptions, it is very challenging to come up with a universal definition of technology. A second difficulty is distinguishing prescientific technology from modern scientific technology. For example, a chalkboard could be considered as a teaching technology in the eighteenth century, but now, other tools such as tablets, learning management systems, and interactive whiteboards have evolved by leveraging technology tools to change the pedagogy of how we teach. The third difficulty is to distinguish technology as equipment or as an applied science. For example, should we identify the computer as technology, or should we identify the computational results produced by a computer as technology. To avoid these concerns, technology scholars argue that people should contextualize technology to human life, related to living in a technological age first, and only then examine technological meaning and understanding without context.

Another approach to defining technology approaches it from a philosophical lens. Tiles and Oberdiek (2013) described technological development as the problem solver and alleviator, but also something that can create more problems as its use becomes part of society. For example, think about smartphones. While this technology has made solving problems, navigating unknown areas, and other information tasks easier, it has created problems with people losing human to human interaction skills. Through different conceptions of technology development and of the relationship between humans and technology, two conflicting views emerged: technological optimism and technological pessimism (Tiles & Oberdiek, 2013). The optimism view of technology states that technology is applied science and can be used to solve problems. In this view, technology and its production have neutral value, general knowledge could be applied in a similar situation. Under optimism, when technology was introduced into social contexts, irresponsible use was due to human choice not the tool itself. Pessimism focuses on the technological system and technical practice more than technology device. Technology practice might dominate and control everything in human life, such as the science, art, culture, economy, and even the ways of doing and making, which shapes humans' social, moral and political life. In the pessimistic view, the human life is influenced by the technology and associated practices. Heidegger (1977) cautioned that technology is something to be thought about and managed by humans and not the other way around (Dreyfus & Spinosa, 2003). This challenge connects to the ethical dilemmas faced by other STEM fields. Engineering often struggles with how innovations may affect lives and how humans will interact. Biologists often struggle with the ethics of medical treatments weighing the benefits versus the risks. The same challenges exist for technology.

Developing a clear definition of technology is complicated by the rapid speed at which technology develops. In the 1960's and 1970's Howard Moore predicted that processing power would double every two years as advances in semiconductor manufacturing increased (Theis & Wong, 2017). As Moore's law accurately predicted technology made exponential leaps as chips could become smaller and more complex. Technology was able to do more things, which created the need to examine if we should do all the things that technology made possible. As technology leaped so did the need to better understanding of what technology is and how it affects our lives. It became more difficult to define what technology is because as it became more complex and more integrated. Our definitions, often based on practical applications, had to radically change. Definitions based on more philosophical examinations of technology retain relevancy as technology evolves, but both require reflection and examination to keep up with the pace of innovation in technology and its application in science, engineering and math.

2.3 Research About Nature of Technology in Education

Many research studies have been conducted to identify the different perceptions about the nature of technology with students and teachers at various levels of education. When searching the literature, the most common conception of technology views technology as an instrument or device. Viewing technology systemically or connected to human practice were barely mentioned (DiGironimo, 2011; Fernandes, Rodrigues, & Ferreira, 2017; Sundqvist & Nilsson, 2018; Waight & Abd-El-Khalick, 2012). The results of these studies show that students and teachers lack nature of technology knowledge. For example, Fernandes et al. (2017) generalize four categories to identify the concepts of the nature of technology, (1) instrumental concepts, which is characterized as tools, artifacts, and machines; (2) cognitive concepts, which is characterized as applying the theoretical knowledge; (3) systemic concepts, which is characterized as the components of a complex system; and (4) value-based concepts, which is characterized as personal value and judgement of science. From a survey and semi-structured interviews of 20 international youth participants, 13 out of 20 students represented the instrumental concepts of technology, nine of the 13 participants focused on the electronic equipment, such as computer, tablet, video games and phones. Five out of 20 students held the cognitive concept of technology, they thought technology is the application of theoretical knowledge. Two participants held the systemic concept of technology, which included the ethical and environmental implication in social context; and three respondents matched with the value-based concept of technology, which was based on participants' personal view. This study illustrated the challenge that the STEM fields face in making use of technology, the different and incomplete views of the nature of technology can hamper collaboration and innovation because all of the team are viewing technology differently.

Sundqvist and Nilsson (2018) surveyed 102 pre-school staff members from Sweden to identify their view about technology education in preschool. The participants confirmed that they emphasized seven categories of technology during their work: "(1) artifacts and systems in children's environment, (2) create, (3) problem solving, (4) concept of technology, (5) the technological experiments, (6) the technique skills, and (7) the natural science" (p.29). The staff members barely taught technology, instead, they provided the materials and created the environment for children and inspired children to experience their world which included technology as part of it. Also, as we can see from this study, much like in other STEM areas, technology views are greatly influence by the sociocultural nature of science. The Swedish views are influenced by their life experiences and cultural approaches to problems.

Concepts of the nature of technology from middle/high school students and teachers were studied by researchers as well. For instance, DiGironimo (2011) developed a framework of the nature of technology, which explained the five dimensions of technology, technology as artifacts, as creation process, and as human practice, history of technology and current role of technology in society. Then the author surveyed 20 middle school students the question: What is technology? and analyzed students' responses according to the framework. The results showed that 50% of students considered technology as artifacts, and only 26.5% students mentioned the current role of technology in society. A mere 2.9% of students thought technology was a human practice, 8.8% students described the history of technology, and 11.8% stated technology as creation process. Waight (2014) addressed 30 science teachers' concepts of the nature of technology through interviews, three major themes emerged as "(1) improves and make life easier, (2) artifacts which function to accomplish tasks, and (3) representations of advances in civilization" (p.1155). According to the study, the science teachers understandably held an optimistic view of technology. But the results indicate another problem, when science teachers exhibited an incomplete view of the nature of technology or a background bias, would they be able to effectively promote the development of the nature of technology in their students? Could they discuss the human, historical and ethical considerations that technology influenced on other science, engineering, and math pursuits? Who bears the responsibility to teach the nature of technology? And where should it exist in the curriculum? In some cases, technology has been included as part of STEM education, but in others, it fits into library studies or business education. To better understand how technology is treated in the curriculum, it is important to understand the standards that many educators use to make sure that they are teaching about technology.

2.4 Standards Movement in Technology Education

Unlike science education where the Nature of Science is included in major standards documents like the Next Generation Science standards, the nature of the technology is not included in major standards movements within educational technology but instead takes a process view about how technology is integrated or applied. Math standards tend to be led by NCTM (National Council for the Teachers of Mathematics). Engineering standards are ITEEA (International Technology and Engineering Educators Association) and IEEE (Institute of Electrical and Electronics Engineers) - which represents the values of the professional organizations for which they prepare students to enter. Other content areas such as TESOL (Teachers of English to Speakers of Other Languages) have their own technology standards as well (2008). In many schools, technology is not an independent subject and is expected to be taught and utilized across the curriculum. In other districts, technology is considered a special topic or something relegated to exploration classes or centers on an infrequent basis. The fact that technology is not a core subject limits both deep learning and the discussion of its implication on society and its integration with other STEM fields.

Technology standards generally begin with a general conception that using technology is in itself good and desirable for teaching and learning and to promote "digital age learning" (ISTE, 2018a, 2018b). Technology integration standards are drafted by International Society for Technology in Education (ISTE). Their standards are broken into audiences. For example, ISTE has specific standards for students in K12 classrooms, educators, and leaders. Technology professionals (coaches, coordinators, etc..) have their own standards as well. Each of these standards mirror each other and focus on how technology is used to promote learning. This organization is directed toward educators who teach with technology and not necessarily technology related careers, so their focus is integration. This focus on integration philosophically focuses on technology as a tool but can lack the other dimensions of the nature of technology.

The ISTE standards focus on the "soft skills" at all levels for students, faculty and administrators. They stress that learners, educators and leaders, use collaboration, data, and communication to use technology for meaningful learning. These consistent references to these twenty-first century skills help to define the nature of technology in the same way that the nature of science is defined by empirical social and cultural relationships, and collaboration (Lederman, 2007). The ISTE standards were first written in 1998, which at that time included a specific section on the "Social, ethical, and human issues" of technology, but these have been removed since the 2007 revision where it has been included as an aspect of digital citizenship (Thomas & Knezek, 2008). Digital Citizenship focuses on how students interact with each other online and often includes a discussion of rules and policies for educators (Hollandsworth, Dowdy, & Donovan, 2011). These concerns mirror the ethical nature of science, however again they focus on the use of technology in education and since the 2007 revision do not specifically address the ethical nature of technology. In later revisions the standards are more general and focus less on content knowledge about technology and more about processes to make effective use of technology to improve learning (ISTE, 2018a, 2018b). Each of the ISTE standards for students, educators, and leaders include digital citizenship. The ISTE standards also lack direct application to other STEM fields. While they mention creativity and collaboration throughout, the connection to science, engineering and math is left for students and teachers to draw themselves.

Research in educational technology tends to follow to the same trends as the standards. The studies are segmented and look at particular soft skills but never really get into deep definitions about the nature of technology. Within educational technology one of the prevalent models for assessing the quality of technology integration and preparation is the TPACK model (Koehler & Mishra, 2009). This model has its roots in the PCK model (Shulman, 1987) which started as a largely science education model that talked about the nature of teaching knowledge. Shulman argued that in order to teach science there is a combination of knowing the scientific content and how to teach specific scientific content. Where the two intertwined was PCK - or specific knowledge about how to teach specific content. Koehler and Mishra (2009) built upon this model to add technology as an equal concern to both content and pedagogy - and created Technological Pedagogy and Content Knowledge (TPACK). Neither of these models examine what is nature of the content knowledge or the nature of technology - instead they look at how technology is taught. One of the criticisms of the TPACK model is the lack of a clear definition of what technology is, and how technology differs from pedagogy. Other criticism includes TPACK's emphasis on technology integration as its own domain away from pedagogy and instead of an examination of the nature of technology, TPACK emphasizes discrete technology knowledge or skills (Parr, Bellis, & Bulfin, 2013). However, given the foundation of TPACK is PCK which has been widely been explored in other STEM fields for decades, it provides an opportunity for exploring the implications of technology use in the teaching of STEM topics as well. Through these shared philosophies, STEM educators can be engaged to look at how technology affects their development of both content knowledge and pedagogy.

Computer science is a growing area of emphasis in the United States and around the world (Code.org, 2018). ISTE has specific standards for computer science educators, and in 2018 released specific standards about content for supporting computational thinking (ISTE, 2018a, 2018b). Computer science standards for K-12 education are relatively new. The Computer Science Teachers Association (CSTA) which is a Division of the Association for Computing Machinery (ACM), the professional organization for computer scientists. The newest version of the CSTA standards were accepted in 2017 and provide the framework for 22 states that have currently adopted them and 11 states that are developing statewide computer science standards (Code.org, 2018). However, when one looks at all of these standards movements, one would not find much examination of what technology is or any epistemological thinking about technology as a way of knowing, There is much attention spent to how to computer scientists interact and do their work. This relates back to the idea that technology is a process or a skill.

The Computer Science Standards developed by CSTA have a similar focus on soft skills but also have more specific knowledge topics, much more like the next generation science standards. They focus on having learners learn how technology is done in the field - career specific processes like collaboration, security, and data protection. However, they have a broader focus on equity and the ethical use of computing and its impact on society. Prior to these standards, there was not much emphasis within technology education on the ethical or social impacts of technology, beyond looking at digital citizenship (Hollandswort, Dowdy, & Donovan, 2011; Lenhart et al., 2011).

One of the ways in which states are measuring their integration of computer science education is by measuring how many students are taking an advanced placement (AP) exam in computer science (Code.org, 2018). As the AP Exam for computer science was being developed, a framework of seven computer science principles were laid out to develop the exam. These principles are show the closest linkage to the development of a nature of technology within an instructional framework (Grover & Pea, 2013 (page 39).

1. Computing is a creative human activity
2. Abstraction reduces information and detail to focus on concepts relevant to understanding and solving problems
3. Data and information facilitate the creation of knowledge
4. Algorithms are tools for developing and expressing solutions to computational problems
5. Programming is a creative process that produces computational artifacts
6. Digital devices, systems, and the networks that interconnect them enable and foster computational approaches to solving problems
7. Computing enables innovation in other fields, including science, social science, humanities, arts, medicine, engineering, and business.

These Computer Science Principles do not seem to be well developed beyond a framework for the AP exam and the CSTA standards instead uses a framework of concepts and competencies. Concepts include computing systems, networks and the internet, data and analysis, algorithms and programming and impacts of computing. They instead list a series of practices that demonstrate computer science. These practices include, fostering an inclusive computing culture, collaborating around computing, recognizing and defining computational problems, developing and using abstractions, creating computational artifacts, testing and refining computational artifacts and communicating about computing (K–12 Computer Science Framework 2016). With the exception of the impacts of computing, these concepts and practices are much more applied and not defining technology or looking at the deeper philosophical issues about computing but instead present best practices without encouraging students to evaluate their inherent value (K–12 Computer Science Framework 2016). The CSTA standards also do not explicitly make connections to other STEM fields and discuss how technology is applied to solve problems in these areas, but instead focuses on computer science as an independent domain.

Computer science education research is another research strand of interest to the nature of technology. This area of research tends to focus on computational thinking, problem formation and solving (Grover & Pea, 2013). This research focuses more on how people learn to think about technology but again does not focus on the nature of technology itself. To further complicate the advancement of this line of thinking, it occurs in multiple locations. While the education research literature on computational thinking is relatively new, it has existed in computer science venues (i.e. ACM journals) for decades (Grover & Pea, 2013). However, fundamental concepts in computer science like abstraction are mirrored in science and mathematics standards and offer a wonderful opportunity to integrate the study of what is technology with the other disciplines while also looking at the impact of technology on the world through the lens of science or math curriculum (Cetin & Dubinsky, 2017; Kramer, 2007).

If we want technology to be a focus of STEM curriculum, we need to learn from other standards developed by organizations where the focus of the standards (science, math, etc.) are philosophically examined as part of the standards. In addition, recent criticism about technology use in education points to the general issue that technology use and integration is not well examined as a social good. It is this kind of orientation that looks at technology without teaching students how to manage it, evaluate its role in their life, examine its relationship to other fields and acknowledge innovation that results. There are a flurry of articles that regularly appear in the mainstream news like Richmond and Troisl (2018) which asks if computers should be banned from the classroom due to things like texting or distraction. This shows that instead of examining the affordances of technology or the application of technology to our lives, the authors are defining technology by device or utility. Another common example seen in popular articles are stories about Silicon Valley parents who will not let their children use technology due to its addictive nature (Bowles, 2018). If we were engaged with discussions about the nature of technology from early ages and explicitly reflecting on it, these discussions would go in a different direction. This lack of examination of what technology is, what are the social benefits and concerns, and what does technology mean to our way of life (i.e. an examination of the nature of technology may be at the root of the debates and concerns. In addition, by looking at the nature of technology, the public would have more tools to assess and participate in innovation and understand how issues like data sharing and privacy can be weighed against the value of available innovations. As both technology and science continue to evolve, these ethical and philosophical questions become key to both an informed electorate and an engaged scientifically literate population.

2.5 Conclusion

The way that we work in the world is changing. The World Economic Forum (2018) reported that many of today's jobs will not exist in 2022, and the main driver of that change is adoption of technologies of like cloud computing, automation, data analytics, and robotics. The report goes on to say that there will be a lack of skilled workers to the do the work. Technological changes expose a greater issue that needs to be addressed, that is the philosophical and ethical nature of technology. While it might be cost effective to rely on technology to do the jobs of the future, are we preparing our citizens to be thoughtful about the nature of technology and how it impacts our society? Are we preparing our workers to be both optimistic about the value that technology brings to our society but also pessimistic about how humans must guard against unintended consequences of technology use? Because technology is necessary part of modern life, having explored the nature of technology can help citizens explore the relationship between technology and themselves, and make informed choices. This understanding and reflection of the nature of technology is integral to the study and advancement of other STEM fields that rely on both the data and collaboration abilities that technology affords.

We are at a pivotal point related to technology education in the United States and throughout the world (Hubwieser et al., 2015). The Summit on STEM Education (White House, 2018) stressed that computer science is key to the future of our economy and is their priority for STEM Education. The impetus of the computer science education movement and adoption of computer science standards around the globe presents an opportunity for educators to incorporate the nature of technology in new meaningful ways and to reflect on how technology affects our lives. Not only can examining the nature of technology help lead to thoughtful and ethical technology use but it can it also help us address issues of equity and access. These issues are shared by all the STEM fields and their discussion offers another opportunity for technology to better contribute to STEM education discussions and advancement.

Understanding the nature of technology and technology itself seems urgent in K-12 education. By mapping the nature of science aspects on to computer science standards and activities, this can be easily seen (Lederman, 2007). For example, the CSTA standards deal specifically with the sociocultural nature of technology by discussing and assessing technology on everyday life. In addition, the standards and computer science education activities focus on access and equity in computer science, and much of the computer science standards movement is driven by this social aim. Activities throughout different computer science curricula illustrate observation and inference especially when it comes to debugging and problem solving in code. Computer programming is a highly creative activity and that is considered an important part of the problem-solving process. These aspects of the nature of technology are very important, but much like the nature of science can only be taught and explored when students are engaged in explicit reflective activities (Lederman, 2007). Webb (2008) discussed the value of technology to traditional science education in that affordances that come with technology like simulations can support student inquiry and argumentation. The nature of technology is compatible with the nature of science, and through the affordances of technology we can engage learners in thinking deeply about the nature of all STEM fields.

To achieve our goals, there is more research to be done. More research could be performed on how people view technology and how they relate it to human endeavors. More research could be designed about technology as a thought process and approach versus practical uses. More research could investigate technology as a way of knowing and solving problems. Also, the research could be expanded to connect technology to other STEM curriculum, such as how to use the nature of technology to enhance student STEM learning and the advancement of STEM content.

References

Berman, F., & Cerf, V. G. (2017). Social and ethical behavior in the internet of things. *Communications of the ACM, 60*(2), 6–7.

Bowles, N. (2018). A dark consensus about screens and kids begins to emerge in Silicon Valley. *The New York Times, 26*.

Cetin, I., & Dubinsky, E. (2017). Reflective abstraction in computational thinking. *The Journal of Mathematical Behavior, 47*, 70–80.

Code.org. (2018). *State of computer science education*. Retrieved from: https://advocacy.code.org/
CSTA. (2017). *Computer science standards*. Retrieved from: https://www.csteachers.org/page/standards
DiGironimo, N. (2011). What is technology? Investigating student conceptions about the nature of technology. *International Journal of Science Education, 33*(10), 1337–1352.
Dreyfus, H. L., & Spinosa, C. (2003). Further reflections on Heidegger, technology, and the everyday. *Bulletin of Science, Technology & Society, 23*(5), 339–349.
Fernandes, G. W. R., Rodrigues, A. M., & Ferreira, C. A. (2017). Conceptions of the nature of science and technology: A study with children and youths in a non-formal science and technology education setting. *Research in Science Education*, 1–36.
Grover, S., & Pea, R. (2013). Computational thinking in K–12: A review of the state of the field. *Educational Researcher, 42*(1), 38–43.
Heidegger, M. (1977). *The question concerning technology* (pp. 3–35). New York: Harper & Row.
Hollandsworth, R., Dowdy, L., & Donovan, J. (2011). Digital citizenship in K-12: It takes a village. *TechTrends, 55*(4), 37–47.
Hubwieser, P., Giannakos, M. N., Berges, M., Brinda, T., Diethelm, I., Magenheim, J., et al. (2015, July). A global snapshot of computer science education in K-12 schools. In *Proceedings of the 2015 ITiCSE on Working Group Reports* (pp. 65–83). ACM.
ISTE. (2018a). *ISTE standards*. Retrieved from: http://www.iste.org/standards
ISTE. (2018b). *ISTE announces new computational thinking standards for all educators standards*. Retrieved from: https://www.iste.org/explore/Press-Releases/ISTE-Announces-New-Computational-Thinking-Standards-for-All-Educators
K–12 Computer Science Framework. (2016). Retrieved from http://www.k12cs.org
Koehler, M., & Mishra, P. (2009). What is technological pedagogical content knowledge (TPACK)? *Contemporary Issues in Technology and Teacher Education, 9*(1), 60–70.
Kramer, J. (2007). Is abstraction the key to computing? *Communications of the ACM, 50*(4), 36–42.
Lederman, N. G. (2007). Nature of science: Past, present, and future. In S. K. Abell & N. G. Lederman (Eds.), *Handbook of research on science education* (pp. 831–880). New York: Routledge.
Lenhart, A., Madden, M., Smith, A., Purcell, K., Zickuhr, K., & Rainie, L. (2011). Teens, kindness and cruelty on social network sites: How American teens navigate the new world of "Digital Citizenship". In *Pew Internet & American Life Project*. Retrieved from: http://www.pewinternet.org/2011/11/09/teens-kindness-and-cruelty-on-social-network-sites/
Mitcham, C. (1994). *Thinking through technology: The path between engineering and philosophy*. Chicago: University of Chicago Press.
Pacey, A. (1983). *The culture of technology*. Cambridge: MIT press.
Parr, G., Bellis, N., & Bulfin, S. (2013). Teaching English teachers for the future: Speaking back to TPACK. *English in Australia, 48*(1), 9.
Richmond, A. S & Troisl, J.D. (2018). *Technology in the classroom: What the research tells us*. Retrieved from: https://www.insidehighered.com/digital-learning/views/2018/12/12/what-research-tells-us-about-using-technology-classroom-opinion
Shulman, L. (1987). Knowledge and teaching: Foundations of the new reform. *Harvard educational review, 57*(1), 1–23.
Sundqvist, P., & Nilsson, T. (2018). Technology education in preschool: Providing opportunities for children to use artifacts and to create. *International Journal of Technology and Design Education, 28*(1), 29–51.
TESOL (2008). *TESOL technology standards framework*. Retrieved from: https://www.tesol.org/docs/default-source/books/bk_technologystandards_framework_721.pdf?sfvrsn=2&sfvrsn=2
Theis, T. N., & Wong, H. S. P. (2017). The end of Moore's law: A new beginning for information technology. *Computing in Science & Engineering, 19*(2), 41.
Thomas, L. G., & Knezek, D. G. (2008). Information, communications, and educational technology standards for students, teachers, and school leaders. In *International handbook of information technology in primary and secondary education* (pp. 333–348). Boston, MA: Springer.

Tiles, M., & Oberdiek, H. (2013). Conflicting visions of technology. In R. C. Scharff & V. Dusek (Eds.), *Philosophy of technology: The technological condition: An anthology* (pp. 249–259). Wiley.

Waight, N. (2014). Technology knowledge: High school science teachers' conception of the nature of technology. *International Journal of Science and Mathematics Education, 12*(5), 1143–1168.

Waight, N., & Abd-El-Khalick, F. (2012). Nature of Technology: Implications for design, development, and enactment of technological tools in school science classrooms. *International Journal of Science Education, 34*(18), 2875–2905.

Webb, M. (2008). Impact of IT on science education. In J. Voogt & G. Knezek (Eds.), *International handbook of information technology in primary and secondary education* (Vol. 20, pp. 133–148). Springer

White House. (2018). *Summit on STEM education.*. Retrieved from: https://www.whitehouse.gov/wp-content/uploads/2018/06/Summary-of-the-2018-White-House-State-Federal-STEM-Education-Summit.pdf

World Economic Forum. (2018, September). *The future of jobs: Global challenge insight report, World Economic Forum*. Geneva: http://reports.weforum.org/future-of-jobs-2018/key-findings/

Zvorikine, A. (1961). The history of technology as a science and as a branch of learning: A Soviet view. *Technology and Culture, 2*(1), 1–4.

Theresa A. Cullen is the Department Head of Curriculum and Instruction at Arkansas Tech University. She was the Director of Digital Strategy and an Associate Professor at the Jeannine Rainbolt College of Education at the University of Oklahoma. She coordinated the undergraduate technology integration courses and the 1-to-1 iPad program for all students studying to be teachers in grades Pre-K to 12. She became an Apple Distinguished Educator in 2015 and is currently the 2020 Research Chair for the ISTE Conference. She earned her PhD in Instructional Systems Technology from Indiana University.

Meize Guo is a Doctoral Candidate in Instructional Systems Technology and minored in Science Education at Indiana University. Her research focuses on technology integration and computer science education, especially on teacher education and professional development. Currently, she is researching elementary STEM teachers' conception and practice of teaching computer science.

Chapter 3
Toward Defining Nature of Engineering in the Next Generation Science Standards Era

Hasan Deniz, Ezgi Yesilyurt, Steven J. Newman, and Erdogan Kaya

3.1 Introduction

The Next Generation Science Standards (NGSS Lead States, 2013) placed a special emphasis on engineering and identified engineering as a discipline to be included in science classrooms in the United States. Engineering secured a higher status in the NGSS compared to the previous science education policy documents (AAAS 1993; NRC 1996). Engineering like science can be conceived as including three domains: (a) engineering as a body of knowledge, (b) engineering as a set of practices, and (c) engineering as a way of knowing (nature of engineering) (Hartman, 2016). It is reasonable to think that there is a parallel between NOE and nature of science (NOS) aspects when we read NOE conceptualizations (Hartman, 2016; Karataş, 2009; NRC, 2012) in light of NOS conceptualizations (Lederman, 2007; Osborne, Collins, Ratcliffe, Millar, & Duschl, 2003). NOS is a well-established research area in science education. Therefore, the developing NOE research agenda and the attempts to conceptualize NOE ideas can be informed by the historical trajectory of NOS scholarship in science education. One can realize that NOE conceptions included in the NGSS are similar to NOS conceptions included in Appendix H of the NGSS and the agreed upon NOS conceptions (Lederman, 2007; Osborne et al., 2003).

Understanding NOS is considered as an essential component of scientific literacy (AAAS 1993; Abd-El-Khalick & Lederman, 2000; NRC, 2000) since it provides an insight into how scientific knowledge develops in real world. Similarly, NOE can be considered as an important component of larger STEM literacy because

H. Deniz (✉) · E. Yesilyurt · E. Kaya
University of Nevada Las Vegas, Las Vegas, NV, USA
e-mail: hasan.deniz@unlv.edu

S. J. Newman
Indiana University, Bloomington, IN, USA

V. L. Akerson, G. A. Buck (eds.), *Critical Questions in STEM Education*, Contemporary Trends and Issues in Science Education 51,
https://doi.org/10.1007/978-3-030-57646-2_3

it helps people to appreciate creative, subjective, tentative, social, and socio-cultural aspects of engineering design process while engineers develop solutions to human problems. In line with other chapter authors in this book, we view STEM as meaningful interdependence among all disciplines of STEM. In other words, includes all individual disciplines of STEM (science, technology, engineering, and mathematics) in a way that is meaningful and showcases the interdependence of the fields. Epistemic beliefs about engineering (NOE views) can influence the knowledge acquisition, interpretation of problems, and the selection of problem-solving strategies during the engineering problem solving (McNeill, Douglass, Koro-Ljungberg, Therriault, & Krause, 2016). From this perspective, NOE should be an integral part of an effective engineering education (Karatas, Goktas, & Bodner, 2010). In this chapter, we aimed to discern NOE aspects relevant to K-12 engineering education from the literature, the Framework (NRC, 2012), and to suggest ways to teach NOE integrated with the engineering design process.

3.2 Nature of Engineering

Unlike NOS, there is no consensus about NOE conceptions relevant to K-12 education (Karataş, Bodner, & Unal, 2016). However, there is a need to establish an agreed-upon list of nature of engineering (NOE) aspects to teach NOE as an integral part of the engineering design process described in the NGSS. One can attempt to create a list of NOE aspects by tapping into NOS literature (Karataş, 2009), conducting a Delphi study (Hartman, 2016) and by surveying state standards (Carr, Bennett IV, & Strobel, 2012).

Karataş (2009) developed a list of NOE aspects based on literature. The list included the notions that (1) engineering solutions are tentative, (2) they require creativity, imagination, and the ability to integrate knowledge from other disciplines, (3) they utilize current scientific and mathematical theories, and learn from previous successes and failures, (4) they require decision making based on certain criteria and constraints as well ethical considerations, (5) they are socially and culturally embedded, (6) they are goal-oriented addressing a specific human need or desire, (7) they require holistic thinking, and (8) they may vary because there might be more than one solution to the same problem.

Hartman (2016) conducted a 3-phase Delphi study to identify NOE aspects relevant to K-12 engineering education. At least 61 subjects participated in first, second, and third round of the study. Participants included 15 K-12 science teachers, 18 K-12 engineering teachers, 17 science education faculty, and 15 engineering education faculty. Hartman (2016) identified eight NOE aspects. These aspects include the notions that (1) there are multiple solutions to the same problem, (2) engineering solutions require creativity, (3) engineering solutions are reached through an iterative engineering design process involving learning from failure, (4) engineering solutions utilize mathematical, visual, and physical modeling, (5) engineering solutions require communication, (6) engineering solutions are evaluated based on certain criteria and constraints, (7) engineering solutions require collaboration and teamwork, and (8) engineering is a unique way of knowing which has similarities and differences with science and other disciplines. These eight NOE aspects were selected based on consensus and stability criteria. Consensus criterion indicates that at least 75% of participants rated each NOE aspect at four or greater on a 5-point scale (4-important, 5-very important) and stability criterion means that for each NOE aspect a similar level of consensus was achieved in both round two and round three. After achieving a consensual list of NOE aspects as a result of a Delphi study, Hartman (2016) suggested major revisions on this list. He recommended to include *the design process* to the list because *the design process* was the second most highly rated NOE aspect in the second round but this particular aspect was not included in the third round due to stability criterion. He also noted that the Delphi study participants warned against teaching the engineering design process as a list of steps similar to the steps of *the so-called scientific method*. Hartman also suggested to treat *creative, communicative, and collaborative* NOE aspects as practices of engineering rather than NOE aspects. Hartman differentiated *unique way of knowing* NOE aspect from other NOE aspects by stating that "this is an overarching concept in engineering and not an aspect of the nature of engineering" p.124). After these major revisions, Hartman's NOE list included five NOE aspects: design driven, divergent, iterative model-driven, and constrained by criteria.

Carr et al. (2012) surveyed all 50 states academic standards to identify instances of engineering content in existing standards to attempt to determine if a consensus on the big ideas of engineering may already exist. Engineering skills and knowledge were found in 41 states' standards. Word analysis was performed on both state engineering and design standards to visually portray the "big ideas" of engineering. The top five "big idea" words identified in engineering standards were: design, technology, use, process, and problem (Carr et al., 2012). The top five verbs prevalent in the design standards were: need, criteria, constraints, model, and data (Carr et al., 2012). When the standards were compiled, an inconclusive consensus of 19 "big ideas" of what consisted of engineering was found. Hence, this further demonstrates that potential NOE aspects have been present in existing state standards, yet never explicitly termed and defined to be included in national standards.

Even though the term nature of engineering (NOE) was not used in the Framework (NRC, 2012) and the NGSS, these documents include NOE ideas throughout if one looks specifically for them. Table 3.1 illustrates the NOE aspects that we discerned from the Framework (NRC, 2012) and the NGSS. Excerpts that are provided in Table 3.1 are by no means exhaustive. We chose to include these particular excerpts because they clearly reflect their corresponding NOE aspect. The italic text in Table 3.1 represents our own definition of NOE aspects.

Table 3.1 Descriptions of Nature of Engineering (NOE) Aspects

NOE Aspect	Description
Demarcation (What is engineering? What makes engineering different from other disciplines?)	*Engineering is systematically engaging in the practice of design to achieve solutions for specific problems. Engineers apply their understanding of the natural world (scientific knowledge) to design solutions for real world problems.*
	In the K-12 context, "science" is generally taken to mean the traditional natural sciences: physics, chemistry, biology, and (more recently) earth, space, and environmental sciences…We use the term "engineering" in a very broad sense to mean any engagement in a systematic practice of design to achieve solutions to particular human problems. Likewise, we broadly use the term technology to include all types of human-made systems and processes-not in the limited sense often in schools that equates technology with modern computational and communications devices. Technologies result when engineers apply their understanding of natural world and of human behavior to design ways to satisfy human needs and wants. (NRC, 2012, p. 11–12)
Engineering design process (EDP)	Engineering design process is viewed as both domain knowledge and the central practice of engineering in the Framework (NRC, 2012). There is an overall agreement on the components of the engineering design process described below.
	The core idea of engineering design includes three component ideas: Define, design, and optimize
	A. Define: Defining and delimiting engineering problems involves stating the problem to be solved as clearly as possible in terms of criteria for success and constraints or limits.
	B. Design: Designing solutions to engineering problems begin with generating a number of possible solutions. These potential solutions are then evaluated to assess which ones best meet the criteria and constraints of the problem.
	C. Optimize: Optimizing the design solution involves a process in which solutions are systematically tested and refined and the final design is improved by trading off less important features for those that are more important (NGSS Lead States, 2013, Appendix I, p. 104)
Empirical basis	*Engineers optimize their design solutions and compare alternative solutions based on evidence obtained from test data.*
	… engineers engage in testing that will contribute data for informing proposed designs. A civil engineer, for example, cannot design a new highway without measuring the terrain and collecting data about the nature of the soil and water flows (NRC, 2012, p. 45)
	Engineers use investigation both to gain data essential for specifying design criteria or parameters and to test their designs. Like scientists, engineers must identify relevant variables, decide how they will be measured, and collect data for analysis. Their investigations help them to identify how effective, efficient, and durable their designs may be under a range of conditions (NRC, 2012, p. 50)

(continued)

Table 3.1 (continued)

NOE Aspect	Description
Tentativeness	*Engineering design solutions are subject to change. Engineering design solutions can be revised to better achieve the desired goal or they can be revised to satisfy different criteria.*
	Phases of engineering design process do not always follow in order, any more than do the "steps" of scientific inquiry. At any phase, a problem solver can redefine the problem or generate new solutions to replace an idea that is just not working out (NGSS Lead States, 2013, Appendix I, page 104).
Creativity	*Creativity and imagination of engineers play a major role during the engineering design process. The role of creativity and imagination is not limited to any specific component of the engineering design process.*
	Engineering and science are similar in that both involve creative processes, and neither uses just one method (NRC, 2012, p. 46).
Subjectivity	*There is no unique solution to an engineering design problem. While there can be many solutions to the same problem, some of these solutions may be more suited to meet the criteria and constraints of the problem.*
	There is usually no single best solution but rather a range of solutions. Which one is the optimal choice depends on the criteria used for making evaluations (NRC, 2012, p. 52).
Social aspects of engineering	*Engineering is not a solitary pursuit. Engineering design solutions are constructed through social negotiation. Despite their individual differences, members of an engineering community share common understandings, traditions, and values. This social dimension enhances the quality of engineering design solutions.*
	The work of engineers, like the work of scientists, involves both individual and cooperative effort; and it requires specialized knowledge (NRC, 2012, p. 28).
	Engineers cannot produce new or improved technologies if the advantages of their designs are not communicated clearly and persuasively. Engineers need to be able to express their ideas, orally and in writing, with the use of tables, graphs, drawings, or models and by engaging in extended discussions with peers. Moreover, as with scientists, they need to be able to derive meaning from colleagues' texts, evaluate the information, and apply it usefully. In engineering and science alike, new technologies are now routinely available that extend the possibilities for collaboration and communication (NRC, 2012, p. 53).

(continued)

Table 3.1 (continued)

NOE Aspect	Description
Social and cultural embeddedness	*Engineering is a human activity. There is a continued interaction between engineering and society. Sociocultural factors influence the engineering design process, and in turn, engineering influences the society. These social and cultural factors include social composition, religion, worldview, political, and economic factors.*
	Not only do science and engineering affect society, but society's decisions (whether made through market forces or political processes) influence the work of scientists and engineers. These decisions sometimes establish goals and priorities for improving or replacing technologies; at other times they set limits, such as in regulating the extraction of raw materials or in setting allowable levels of pollution from mining, farming, and industry (NRC, 2012, p. 213)
Criteria or constraints	*Engineers make decisions based on predetermined criteria and constraints during the engineering design process. Making gains on one criterion often involves losing on another criterion or other criteria.*
	Engineers must contend with a variety of limitations, or constraints, when they engage in design. Constraints, which frame the salient conditions under which the problem must be solved, may be physical, economic, legal, political, social, ethical, aesthetic, or related to time and place. In terms of quantitative measurements, constraints may include limits on cost, size, weight, or performance, for example. And although constraints place restrictions on a design not all of them are permanent or absolute (NRC, 2012, p. 205).
Failure-laden	*Failure in engineering design is inevitable and provides opportunities for improving design solutions. Engineers engage in multiple iterations to enhance the quality of their products and systems.*
	Tests are often designed to identify failure points or difficulties, which suggest the elements of the design that need to be improved (NRC, 2012, p. 207).
	Engineers often look for and analyze patterns, too. For example, they may diagnose patterns of failure of a designed system under test in order to improve the design (NRC, 2012, p. 86).

3.3 Research on NOE

The number of studies examining students' NOE views is limited (e.g., Capobianco, Diefes-Dux, Mena, & Weller, 2011; Fralick et al., 2009; Karataş et al., 2016). The research on NOE views focused mostly on examining students' knowledge about engineering and engineers' work. The Draw an Engineer Test (DAET), which is a modified version of the Draw a Scientist Test (DAST), has been commonly used by the researchers with regard to students' views of engineering (Knight & Cunningham, 2004). This instrument asks learners to draw an engineer at work and provide explanations for their drawings. Studies using DAET illustrated that students had limited knowledge about engineering. They usually associated engineering with building and fixing, and engineers with workers and laborers. Knight and Cunningham

(2004), for instance, studying with 384 K-12 students found that most of the students drew artifacts of the building and fixing such as heavy machinery and hard hat, and products of engineering such as cars, machines and engines. The students' drawings illustrated that students perceived engineers as construction workers or mechanics. Only a few students portrayed engineers thinking and sketching their designs on a paper.

Likewise, Fralick, Kearn, Thompson, and Lyons (2009) attempted to compare and contrast the 1600 middle school students' conceptions of engineers and scientists through DAST and DAET instruments. Students' drawings showed that students perceived engineering as a profession mainly involving physical labor. Specifically, they drew engineers as a doer or a worker bee. They portrayed engineers wearing laborers' clothing in outdoor and operating vehicles, or building structures. Compared to their drawings of scientists, most students depicted engineers with no inferred action which illustrated that students lack knowledge about the engineering design process. Further, students portrayed engineers less scholarly than scientists. While they drew scientists thinking and with books, and mathematical symbols, they portrayed engineers mostly with mechanics tools. All in all, the students had more informed views with respect to the field of science than they do about engineering. They had insufficient knowledge about the cognitive and creative aspects of the engineering design process.

Similarly, Capobianco et al. (2011) interviewed elementary students from each grade by using the DAET instrument to explore their conceptions of engineers. Students' drawings and interview analyses revealed four emergent themes with respect to students' conceptions about engineers. First, students depicted engineers primarily as mechanics who work on vehicles. Several younger students portrayed engineers as an object and an engine. Second, several students perceived engineers as tradesmen who fix and build roads and objects. Third, students depicted technicians who work on electronic devices (e.g., computer, television, telephones). These students thought that the engineering profession requires technical skills. Last, few students, only fourth and fifth-grade students, conceived that the engineering profession involves designing buildings, electronic devices, and vehicles. Although few upper elementary students seemed to have relatively more informed views as compared to lower elementary students, the majority of the fourth and fifth-grade students still perceived engineering as involving mainly mechanical labor. In light of their findings, the researchers made a list of key attributes of an engineer for engineering education. These attributes included creativity, working in teams, using science, mathematics and technology, solving human problems, and designing everything around us.

Although the DAET instrument allows us to capture students' general ideas about engineers and engineering, it does not provide detailed information with regard to students' NOE conceptions across NOE aspects that we described in Table 3.1. For this reason, Karatas, Micklos and Bodner (2011) conducted interview sessions using pictures of engineering artifacts along with sixth grade students' own drawings to assess their NOE conceptions with regard to the definition of engineering, the engineering design process, the demarcation between science and engineering, and the role of engineering in society. Although students in this study thought that building structures, fixing and assembling vehicles as part of the design process, nearly half of the students perceived engineering as an active and dynamic process including planning, designing, creating and testing phases. However, none of the students addressed all phases of the design process and they focused only on two or three phases of the engineering design process. Therefore, the study indicated that students envisioned engineers as skilled craftsmen. As for the demarcation between science and engineering, most of the students struggled to differentiate between science and engineering while only a few students stressed that engineers design and build products while scientists work on nature.

Later, Karataş et al. (2016) developed a 12-item views of the nature of engineering (VNOE) questionnaire to explore the first-year engineering students' views engineering and design process. The inductive data analyses indicated that majority of students perceived engineering profession involving the mental tasks including sketching, conceptual representations of the design, and overseeing the construction. Most of the students thought that engineering involves problem-solving process which aims at improving engineering designs and solving human and environmental problems. On the other hand, students failed to explain the differences between science and engineering and to mention the role of science and math knowledge, and teamwork in engineering design. This study showed the importance of using questions specifically addressing NOE aspects to provide a comprehensive analysis of students' views of NOE. However, the authors did not address the validity and reliability issues regarding the instrument.

Overall, studies indicated that K-12 students held uninformed views of NOE aspects. On the other hand, no valid and reliable instrument is available in the related literature to fully capture K-12 students' NOE conceptions. It is, therefore, necessary to develop a consensus list of NOE aspects relevant to K-12 education, which can pave the way for the development of valid and reliable instruments to assess learners' NOE views.

3.4 Teaching NOE

After the release of the NGSS, mounting pressure has been applied to in-service and pre-service teachers to be able to incorporate both science and engineering into lesson plans via cross-cutting concepts. As highlighted through a recent editorial piece (Akerson et al., 2018), concerns over how to get current in-service and pre-service science teachers to become proficient, not only in teaching NOS, but furthermore NOE, when science educators have not been able to successfully help (or at least most) K-12 science teachers conceptualize NOS, their home discipline (Akerson et al., 2018). This presents a compounding problem in the teacher preparation pipeline—the science teacher educators with no engineering backgrounds are preparing the science teachers, who also have no engineering backgrounds (Akerson et al., 2018). Despite this dire situation, science educators can model teaching NOE in professional development programs and science teaching methods courses by considering effective features of explicit-reflective NOS teaching (Adibelli-Sahin & Deniz, 2017).

Rich NOS literature can offer insights about how to teach NOE aspects as well as NOS aspects. Students should experience the engineering design process in their science classes but the activity itself should not be enough. Students should reflect on the epistemological aspects of the engineering design process while they are actively engaged in the engineering design activity. To this end, the science and engineering education community should achieve a consensus list of NOE aspects relevant to K-12 education and decide which NOE aspects will be taught in each grade level at what level of sophistication similar to the NOS aspects presented in Appendix H of the Next Generation Science Standards.

The research with regard to what kinds of strategies are effective in improving students' NOE views is limited. However, NOS literature is rich with studies suggesting that engaging students in inquiry activities does not necessarily improve their NOS views if the students are not intentionally introduced to NOS ideas through explicit-reflective NOS instruction (Lederman, 2007). Explicit-reflective NOE instruction should intentionally draw students' attention to relevant NOE aspects during the engineering design activity and encourage students to revise their NOE ideas in light of their experience during the engineering design activity. While explicit part of this instruction refers to making NOE aspects visible to the students, the reflective part refers to encouraging learners to revise their NOE ideas in light of their engineering design experiences. For example, Deniz, Yesilyurt, and Kaya (in press) provided a practical example of how explicit-reflective instruction can be adapted to teach NOE aspects. Authors described in detail how they used picture books to teach selected NOE aspects to grades 3–5 elementary students using an explicit-reflective instruction embedded within an engineering design activity.

The authors aligned each phase of the engineering design process with a particular picture book. The authors described how they used picture books to reflect on students' experience in a particular engineering design process phase from the perspective of targeted NOE ideas by the book reading. Each book targeted minimum one and maximum three NOE ideas. For example, during create phase of the engineering design process, the authors read the picture book *The Most Magnificent Thing* (Spires, 2014). This book tells the story of a girl building a special scooter. The picture book conveys the idea that engineers use their creativity and imagination, and should be persistent in the face of failure to successfully complete their projects. By reading this picture book the authors explicitly stated that engineers could revise their design ideas (tentative NOE) by using their creativity and imagination (creative NOE) while persevering in the face of failure (failure-laden NOE). The authors also supported their explicit NOE instruction by referring to a NOE poster strategically placed in the classroom. The authors reminded students to refer to the NOE poster during the engineering design process and they used the poster to explicitly identify NOE aspects during the reflective debriefing activity at the end of the engineering design challenge. The debriefing activity involved specifically prepared questions addressing each targeted NOE aspect. For example, authors asked questions such as "In what ways did you use your creativity and imagination during the engineering design process?" and "Did each group develop the same design?" to start the reflection around creative NOE aspect and subjective NOE aspect, respectively.

3.5 Conclusion

Nature of engineering research is at its infancy. In order to advance NOE as a new research agenda, we need to make progress on three fronts: (a) achieving consensus around NOE aspects relevant to K-12 education, (b) developing valid and reliable instruments to assess students' and teachers' NOE views, and (c) developing effective methods to teach NOE. In order to achieve the same level of rigor in NOS research, achieving consensus around NOE aspects, and then developing valid and reliable instruments to assess these agreed-upon NOE aspects are of primary importance. There were some initial attempts to achieve consensus on NOE engineering ideas relevant to K-12 education (Hartman, 2016; Karataş, 2009), but these attempts still did not establish a secure base for NOE aspects similar to agreed-upon NOS aspects (Lederman, 2007). Recently, the Framework (NRC, 2012) and the Next Generation Science Standards (NGSS Lead States, 2013) portrayed certain NOE ideas without defining NOE as a construct similar to NOS portrayed in Appendix H of the NGSS. Having these valid and reliable instruments can help us to develop more effective methods to teach NOE. The current NOE research primarily uses interviews and Draw an Engineer Test (DAET) to assess learners' NOE views. Nature of engineering needs valid and reliable open-ended instruments similar to the Views of Nature of Science (VNOS) instruments (Lederman, Abd-El-Khalick, Bell, & Schwartz, 2002).

Thanks to the rich NOS literature, we have a significant number of research articles reporting about which instructional approaches are more effective in improving learners' NOS views (Lederman, 2007). We also have a significant number of practitioner articles (Lederman, Bartels, Lederman, & Gnanakkan, 2014; McComas, 2004) and book chapters (McComas, 1998; McComas, 2020) describing effective methods to teach NOS in K-12 education. Fledgling NOE scholarship can easily borrow from this rich NOS literature to design effective teaching strategies in improving learners' NOE views. For instance, we should not expect learners to improve their NOE views simply because they participated in an engineering design activity. The NOS literature is filled with studies reporting that learners may not improve their NOS views even if they participate in a scientific inquiry (Abd-El-Khalick & Lederman, 2000; Bell, Blair, Crawford, & Lederman, 2003).

Engineering literacy has been defined as (a) discuss, critique, and make decisions about national, local, and personal issues that involve engineering solutions, (b) understand and explain how basic societal needs (e.g., water, food, and energy) are processed, produced, and transported, (c) solve basic problems faced in everyday life by employing concepts and models of science, technology, and mathematics (Chae, 2010). We believe that engineering design process along with its epistemological aspects (NOE) should be integrated to science curriculum at each grade level as it was recommended by the NGSS. This integration can be successfully achieved in both elementary (Deniz, Kaya, & Yesilyurt, 2018) and secondary grades (Roehrig, Moore, Wang, & Park, 2012). Through a more integrated approach, students can be exposed to engineering earlier and more, thereby increasing their own engineering literacy.

References

Abd-El-Khalick, F., & Lederman, N. G. (2000). Improving science teachers' conceptions of the nature of science: A critical review of the literature. *International Journal of Science Education, 22*(7), 665–701.

Adibelli-Sahin, E., & Deniz, H. (2017). Elementary teachers' perceptions about the effective features of explicit reflective nature of science instruction. *International Journal of Science Education, 39*(6), 761–790.

Akerson, V. L., Burgess, A., Gerber, A., Guo, M., Khan, T. A., & Newman, S. (2018). Disentangling the meaning of STEM: Implications for science education and science teacher education. *Journal of Science Teacher Education, 29*(1), 1–8.

American Association for the Advancement of Science. (1993). *Benchmarks for science literacy: Project 2061*. New York: Oxford University Press.

Bell, R. L., Blair, L. M., Crawford, B. A., & Lederman, N. G. (2003). Just do it? Impact of a science apprenticeship program on high school students' understanding of the nature of science and scientific inquiry. *Journal of Research in Science Teaching, 40*(5), 487–509.

Capobianco, B. M., Diefes-Dux, H. A., Mena, I., & Weller, J. (2011). What is an engineer? Implications of elementary school student conceptions for engineering education. *Journal of Engineering Education, 100*(2), 304–328.

Carr, R. L., Bennett IV, L. D., & Strobel, J. (2012). Engineering in the K-12 STEM standards of the 50 US states: An analysis of presence and extent. *Journal of Engineering Education, 101*(3), 539–564.

Chae, Y. (2010). AC 2010-1287: Core concepts for engineering literacy: The interrelationship among STEM disciplines. *Age, 15*(1).

Deniz, H., Kaya, E., & Yesilyurt, E. (2018). The soda can crusher challenge: Exposing elementary students to the engineering design process. *Science & Children, 56*(2), 74–78.

Deniz, H., Yesilyurt, E., & Kaya, E. (in press). Teaching nature of engineering with picture books. *Science & Children.*

Fralick, B., Kearn, J., Thompson, S., & Lyons, J. (2009). How middle schoolers draw engineers and scientists. *Journal of Science Education and Technology, 18*(1), 60–73.

Hartman, B. D. (2016). *Aspects of the nature of engineering for K-12 science education: A Delphi study* [Doctoral Dissertation, Oregon State University]. https://ir.library.oregonstate.edu/concern/graduate_thesis_or_dissertations/t148fk693.

Karataş, F. Ö. (2009). *First-year engineering students' views of engineering* (Doctoral dissertation). Retrieved from Proquest.

Karataş, F. Ö., Bodner, G. M., & Unal, S. (2016). First-year engineering students' views of the nature of engineering: Implications for engineering programmes. *European Journal of Engineering Education, 41*(1), 1–22.

Karatas, F.O., Goktas, Y., & Bodner, G. M. (2010). An argument about nature of engineering (NOE) and placing the NOE into engineering education curriculum. In *Proceedings of Turkey's vision 2023 conference series: International engineering education conference*, 4–6 November, Antalya, Turkey.

Karatas, F. O., Micklos, A., & Bodner, G. M. (2011). Sixth-grade Students' views of the nature of engineering and images of engineers. *Journal of Science Education and Technology, 20*(2), 123–135.

Knight, M., & Cunningham, C. (2004, June). Draw an engineer test (DAET): Development of a tool to investigate students' ideas about engineers and engineering. In *ASEE annual conference and exposition* (Vol. 2004).

Lederman, J., Bartels, S., Lederman, N., & Gnanakkan, D. (2014). Demystifying nature of science. *Science and Children, 52*(1), 40–45.

Lederman, N. G. (2007). Nature of science: Past, present, and future. In S. K. Abell & N. G. Lederman (Eds.), *Handbook of research on science education* (pp. 831–879). Mahwah, NJ: Erlbaum.
Lederman, N. G., Abd-El-Khalick, F., Bell, R. L., & Schwartz, R. S. (2002). Views of nature of science questionnaire: Toward valid and meaningful assessment of learners' conceptions of nature of science. *Journal of Research in Science Teaching, 39*(6), 497–521.
McComas, W. F. (1998). *The nature of science in science education: Rationales and strategies*. Dordrecht, The Netherlands: Kluwer Academic.
McComas, W. F. (2004). Keys to teaching nature of science. *Science Teacher, 71*(9), 24–27.
McComas, W. F. (2020). *Nature of science in science instruction: Rationales and strategies* Springer.
McNeill, N. J., Douglass, E. P., Koro-Ljunberg, M., Therriault, D., & Krause, I. (2016). Undergraduate students' beliefs about engineering problem solving. *Journal of Engineering Education, 105*(4), 560–584.
National Research Council. (1996). *National science education standards*. Washington, DC: National Academy Press.
National Research Council. (2000). *Inquiry and the national science education standards*. Washington, DC: National Academy Press.
National Research Council. (2012). *A framework for K-12 science education: Practices, crosscutting concepts, and Core ideas*. Committee on a conceptual framework for new K-12 science education standards. Board on science education, division of behavioral and social sciences and education. Washington, DC: The National Academies.
NGSS Lead States. (2013). *Next generation science standards: For states, by states*. Washington, DC: The National Academies Press.
Osborne, J. F., Collins, S., Ratcliffe, M., Millar, R., & Duschl, R. (2003). What 'ideas about science' should be taught in school science? A Delphi study of the 'expert' community. *Journal of Research in Science Teaching, 40*(7), 692–720.
Roehrig, G. H., Moore, T. J., Wang, H. H., & Park, M. S. (2012). Is adding the E enough? Investigating the impact of K-12 engineering standards on the implementation of STEM integration. *School Science and Mathematics, 112*(1), 31–44.
Spires, A. (2014). *The most magnificent thing*. Tonawanda, NY: Kids Can Press.

Hasan Deniz is a Professor of Science Education at University of Nevada Las Vegas. His research focuses on preservice and inservice teachers' ideas about Nature of Science-Nature of Engineering, Engineering Education, and Evolution Education. He is the Director of Center for Math, Science, and Engineering Education at University of Nevada Las Vegas.

Ezgi Yesilyurt is an Assistant Professor of Life Science Education at Weber State University. Her research focuses on preservice and inservice teachers' ideas about Nature of Engineering, Engineering Education, and Evolution Education.

Steven J. Newman is a PhD student in Science Education at Indiana University. He transitioned from a background in scientific bench research to further education in teaching science at the college level. His research interests are on examining the interaction of co-teaching on teacher knowledge and student-teacher interactions.

Erdogan Kaya is an Assistant Professor of Computer Science Education at George Mason University. His research focuses on preservice and inservice teachers' ideas about Nature of Engineering, Engineering Education, and Computer Science education.

Chapter 4
The Nature of Mathematics and Its Impact on K-12 Education

Rick A. Hudson, Mark A. Creager, Angela Burgess, and Alex Gerber

4.1 Introduction

What is mathematics? This question elicits controversy among both mathematicians and the general public alike. Although mathematical content is accepted as an important component of the K-12 curriculum, what students should learn about the nature of mathematics is not well defined, and the experiences K-12 students have in learning about the nature of mathematics varies greatly.

These differences in students' experiences have an impact on how students develop conceptions about what mathematics is and how it is valued. Boaler (2015) asserts that students' understanding of mathematics is different than their understanding of other fields. She states that children describe science and English in ways that are similar to how professors in these fields describe the subject. However, children view mathematics as focused on numbers or "a set of rules." In contrast, mathematicians often describe mathematics as "a study of patterns." Although much of the K-12 school curriculum focuses on the study of numbers and their properties, ironically a great many mathematicians are not concerned with numbers at all. Such great differences between the mathematics experienced by students and professional mathematicians may help us to account for the lack of interest that many students have in mathematical study.

Mathematicians often classify mathematics into two separate, but complementary fields – pure mathematics and applied mathematics. Pure mathematics is concerned with mathematics for its own sake, whereas applied mathematics seeks to

R. A. Hudson (✉) · M. A. Creager
University of Southern Indiana, Evansville, IN, USA
e-mail: rhudson@usi.edu

A. Burgess · A. Gerber
Indiana University, Bloomington, IN, USA

V. L. Akerson, G. A. Buck (eds.), *Critical Questions in STEM Education*,
Contemporary Trends and Issues in Science Education 51,
https://doi.org/10.1007/978-3-030-57646-2_4

use mathematics at the service of other disciples, such as biology, physics, or finance. These two separate branches of the field of mathematics contribute further to misunderstandings about what the nature of mathematics is and its relationship to STEM. A commonality between the two branches is that both pure and applied mathematicians seek to use reasoning to solve problems. However, the branch of applied mathematics is most often associated with the definition of STEM because applied mathematics displays an interdependence on the contexts in which it is situated, including science, technology, and engineering.

In this chapter, we will describe some of the philosophical underpinnings that motivate the nature of mathematics. We will follow with a discussion of how students experience the nature of mathematics in today's K-12 classroom and provide suggestions about what K-12 students should learn about the nature of mathematics and the role that STEM may have in progressing this discussion.

4.2 Philosophical Underpinnings: Contrasting Absolutist and Fallibilist Perspectives

Although other theories about the nature of mathematics exist, two contrasting perspectives regarding the nature of mathematics are the fallibilist and absolutist viewpoints. The absolutists believe that mathematical knowledge is certain and unchallengeable (Davison & Mitchell, 2008). Furthermore, an absolutist views mathematics as a construct that has always been present and humans just discovered it. In other words, they view math as a divine gift that never has error or contradiction (White-Fredette, 2010). Alternatively, individuals who hold a fallibilist philosophy view mathematics as a human construct, and therefore, as susceptible to falsifiability as any human endeavor may be. Fallibilists believe that mathematics is built upon the needs of the society and limited by cultural boundaries that dictate its certainty and applicability (Hersh, 1997). It should come as no surprise that these two distinct philosophies manifest themselves in mathematics teaching in different ways.

Educators with an absolutist view often see mathematics as a set of rules and procedures that are valuable for solving the problems. These teachers often work to break large ideas down into their most elementary ideas and then construct connections between these elementary ideas that build the larger idea. However, it is often argued that teaching in this way causes students to miss out on opportunities for critical thinking and exploration of the mathematical concepts, because students are simply applying ready-made formulas or procedures often without thought or reason (White-Fredette, 2010).

From an ontological standpoint, fallibilists are more likely to view reality as socially constructed, and therefore hold epistemologies supportive of social theories of knowledge and teaching (Ernest, 2018). Teaching mathematics as a human construct encourages students to invoke critical thinking, socially construct mathematical knowledge and, explore and investigate concepts using mathematical inquiry

(White-Fredette, 2010). Critics of this philosophy often argue that it is wasteful of the precious little time teachers have to force students to "re-invent/discover" the mathematics that took society centuries or millennium to discover. Moreover, teachers will argue that the mathematics that mathematicians do vary from what students will do, and although mathematicians' conjectures are certainly tentative, the theorems K-12 students typically encounter are correct and beyond falsification. This concern is common enough to warrant an example of what is meant by the tentative nature of mathematics before diving too deeply into the implications for teaching this philosophical debate has.

Consider a conversation among a group of third graders who are exploring the concept of multiplication where they have developed an understanding of $A \times B$ as the total number of objects when there are A groups with B objects in each group.

Child 1: Multiplication will always make a bigger number.
Child 2: That's interesting. Let's check 2 times 3. I get 6, which is bigger.
Child 3: I tried 1 times 3 and got 3, which isn't bigger.
Child 1: It will work with all numbers except 1.

This child's conjecture displays thinking that a falibilist would value, as mathematicians regularly look for patterns and make conjectures. However, this student's conjecture is certainly false as many numbers, including 1, refute the conjecture (e.g., zero, negative numbers, and real numbers between 0 and 1). Adults regularly make claims about all numbers when they really mean counting numbers, but more importantly, third graders are likely to not have learned about operations with negative numbers or fractions yet. So, this child's conjecture, if modified to exclude 1, is true in the reality of numbers to that child. The child's conjecture fails because the claim is for all numbers, but the conjecture can be made to be correct, by properly restricting the domain of numbers the conjecture covers (e.g., real numbers greater than 1). In describing the nature of mathematics as tentative, we do not mean to infer that 3 times 2 will someday equal something other than 6. Instead, like in the case of the child's conjecture, what makes mathematics potentially false is that all theorems in mathematics are based on either assumed to be true statements or statements proven to be true. It is through these assumed and foundational statements, like our fictional child's understanding of the concept of number, that the potential for falsity arises. In fact, the mathematical historian, Lakatos (1976) suggested that by wrestling with terms like multiplication, we gain a better understanding of what students mean and in turn create more advanced mathematics.

Having clarified these two contrasting arguments of what is mathematics, we turn to describing the impact of these philosophies on mathematics education, which is an important, yet neglected aspect of math education research. There is little evidence in the literature about what philosophical standpoints mathematics educators hold and how they accommodate curriculum expectations, teaching practices and their own math epistemologies. The philosophy of mathematics is rarely examined by researchers and it is important to discern the understanding of the basic principles and concepts that teachers hold before sustainable reform can be initiated.

Despite prominent philosophers of mathematics education (e.g., Ernest, 2018) advocating that a dichotomy exists between fallibilist and absolutist philosophies of mathematics, others suggest that there is no such sharp distinction and that teachers' philosophies of math run along a continuum with fallibilism and absolutism lying at the extremities (Davison & Mitchell, 2008). An important question to ask is: are teachers even thinking about the nature of mathematics when determining how to teach a given topic? Are they consciously making decisions about the global practices and viewpoints that students will develop about mathematics, or are they looking locally at meeting the content objectives within a single lesson?

It is important to draw a distinction between a teacher's philosophy of teaching and learning and a teacher's philosophy of mathematics. Simply because a teacher believes that knowledge can be appropriated via social interaction, it does not necessarily follow that they hold a fallibilist view that mathematics is a socially-constructed, tentative discipline. Equally, a teacher who employs traditional procedurally-focused instructional strategies may do so, not because they hold an absolutist philosophy, but because they believe this is the most effective strategy to help their students learn specific mathematical content.

4.3 What Should K-12 Students Know About the Nature of Mathematics?

What K-12 students should come to understand about the nature of mathematics is not clearly defined. Although research has offered little advice on how the absolutist and fallibilist philosophies impact teachers' decisions, several curriculum efforts have sought to describe what mathematical practices and processes students should come to enact when doing mathematics. For example, NCTM's *Principles and Standards for School Mathematics* described five process standards to "highlight ways of acquiring and using [mathematical] content knowledge" (2000, p. 29). These processes included problem solving, reasoning and proof, communication, connections, and representations.

These processes and similar mathematical practices have been incorporated into curriculum documents in a number of countries. For example, the Australian Curriculum incorporates the key ideas of understanding, fluency, problem-solving, and reasoning (Australian Curriculum, Assessment, and Reporting Authority, n.d.). The Singapore Mathematics Framework identifies mathematical problem solving as its central focus, but also emphasizes other mathematical processes, including reasoning, communication, and modeling (Ministry of Education, Singapore, 2012). In the United States, the Common Core State Standards for mathematics [CCSSM] (NGA/CCSSO, 2010) include the *Standards for Mathematical Practice* (SMPs), which "describe varieties of expertise that mathematics educators at all levels should seek to develop in their students." For example, students should be able to "make sense of problems and preserve in solving them," which they describe as having a variety of skills like analyzing "givens, constraints, relationships, and goals." The SMPs require students to be the active agents in doing mathematics. Although a teacher can tell students the Pythagorean Theorem, they cannot tell students how to build a logical progression of statements to prove such a theorem. A teacher may demonstrate a proof of the Pythagorean Theorem, but doing so does not guarantee her students will learn how to construct proofs. In fact, the literature is filled with evidence of students at all levels of mathematics achievement who struggle to construct even basic proofs (Healy & Hoyles, 2000; Stylianides, Stylianides, & Weber, 2017).

In the sections that follow, we will describe three fundamental ideas about mathematics that we believe K-12 students should come to recognize about mathematics. These descriptions transcend the mathematical processes and practices described above which focus on students' ability to engage in the work of mathematics. These three ideas describe aspects of mathematics that will support students' understanding of the nature of the field.

4.4 Mathematics Is a Way of Knowing

When approaching a problem from an absolutist perspective, it might seem as if one simply needs to select the appropriate mathematical tool for the job. For example, if one were working a problem involving a right triangle, perhaps trigonometry or the Pythagorean Theorem will be helpful. Absolutists might assume that students must first master procedures, like applying the Pythagorean Theorem to find the length of a hypotenuse, and after doing so they can apply their skills in problem solving situations. By doing this, teachers hope to instill how to reason in mathematics. This absolutist perspective has guided the development of many textbooks where students repeat a procedure multiple times, and then use this procedure to solve a small number of word problems. However, problem-solving situations are rarely this straightforward. In the right triangle example, there are so many options for tools from trigonometry that selecting an appropriate tool is far from a trivial task (Duval, 2006; Weber, 2001). Unfortunately, reasoning is not like the procedural knowledge one finds in traditional textbooks, because there is no pre-determined procedure for how to reason in mathematics. This suggests that reasoning cannot be taught directly. Not shockingly, performance on problems involving reasoning are among the poorest on large-scale national and international assessments (Roach, Creager & Eker, 2016).

From a fallibilist perspective, students need to be encultured to what it means to reason in mathematics. Namely, students need to be given problem-solving tasks, but solving those tasks is not the end goal. Instead the goal is making the processes used explicit when solving the problem, so students might learn what constitutes reasoning in mathematics. Today, many curriculum materials are questioning the popular belief that students must first learn a set of procedures before applying them to solve problems. In one research study, Nathan and Petrosino (2003) asked 48 pre-service high school teachers to rank six related algebra tasks by level of difficulty. One of the ways the tasks varied is that some were purely algebraic (i.e., solve this equation) while others were contextually-based problems (i.e., a story problem that required the student to solve an equation). Some tasks were similar in that they involved the same numbers and operations and solving them produced the same answer. What they found is that most teachers felt that the contextually-based problems would be the hardest for their students because the teachers felt their students would have to first write an equation for the contextually-based problem. This equation was the same equation as the purely algebraic problem, however they felt the contextually-based problem required two steps whereas the purely algebraic problem only required one. This seems logical, however research into how students solve these problems have consistently showed that the students were more likely to solve the contextually-based problem correctly (Nathan & Petrosino, 2003). This is because they did not set up an equation like their teachers imagined. They instead reasoned about the context to help them solve the problem. Studies like these and others provide support for the National Council of Teachers of Mathematics (NCTM) who proposed that procedural fluency should be built from conceptual understanding (NCTM, 2014). Similarly, education researchers have also examined ways that skills and procedures can be co-developed in students with processes like reasoning (Kobiela & Lehrer, 2015).

Like defining what mathematics is, defining reasoning in mathematics is a similarly difficult task. Many education researchers have made efforts to do so (e.g., Jeannotte & Kieran, 2017; Russell, 1999). A summary of these definitions described reasoning as a set of interrelated practices that include generalizing/conjecturing, investigating why, and proving/refuting (Lannin, Ellis, & Elliott, 2011). Although it may appear that these processes happen linearly as they are written, it is often not the case. Instead, it is important for students to be able to participate in the zig-zag nature of mathematical reasoning. This is more aligned with Lakatos's (1976) description of how mathematics developed historically. Although proof might be thought of as the culminating activity of mathematics, and in many cases this is true, Lakatos noted that refutations play a far greater role in mathematical discoveries than proofs. Although teachers might be concerned about a child being wrong, it can be argued that being wrong positions the child to learn a more important lesson that reasoning is the foundation of mathematical ideas.

4.5 Mathematics Is Tentative, Because Mathematical Claims Are Based on Assumptions

One of the challenges of helping students to appreciate the tentative nature of mathematics is that much of the mathematical content that is the focus of K-12 mathematics has been settled among mathematicians for centuries, if not millennia. In contrast, the public regularly has opportunities to understand the tentative nature of science when a new experimental drug shows promise in curing a chronic disease or a new plant hybrid is used to boost crop production. However, the groundbreaking work that is happening in mathematics today is rarely featured in the media, unless someone calculates a few more digits of pi. Mathematical research, especially in pure mathematics, is often inaccessible to an undergraduate student studying mathematics, let alone a member of the general public. As a result, there are fewer opportunities for the public to understand the tentative nature of mathematics, which may be one reason the absolutist views of mathematics are perpetuated.

In mathematics, axioms are described as statements that are assumed to be true. Such assumptions underlie the tentative nature of mathematics. In pure mathematics, axioms are the fundamental ideas with which we build mathematical proofs. Throughout the history of mathematics, there have been a number of discoveries that have occurred as a result of questioning assumptions made. For example, for more than 2000 years, Euclid's five postulates formed the basis of geometric thinking. Euclid's fifth postulate is equivalent to the statement "Given a line and point not on it, at most one line parallel to the given line can be drawn through the point." Although this statement was commonly agreed upon, during the nineteenth century, mathematicians began to assume it was false and investigated the results. Consequently, new Non-Euclidean geometries, such as hyperbolic and elliptic geometries were introduced. Such moments in the history of mathematics provide arguments for the fallibilist perspective.

The commonly cited van Hiele (1985) levels describe students' typical progression of thinking in geometry, and the highest level describes students' ability to compare axiomatic systems. Few students encounter Non-Euclidean geometries before postsecondary mathematics, however K-12 students should come to recognize that one's mathematical claims must be rooted in the assumptions that they make. For example, in creating mathematical models of complex situations, students should reflect on the choices and assumptions that they make throughout the modeling cycle.

How do teachers help students understand the tentative nature of mathematics? Drawing connections to the history of mathematics and their personal explorations of numbers may help to support this understanding. Throughout the K-12 curriculum, students' access to the number system gradually expands. Initially students focus on counting numbers, but as they progress, they develop their view of numbers to include rational, irrational, and eventually complex numbers. This sequence corresponds to how number systems developed historically. Ancient Egyptians used unit fractions to describe rational numbers. (For example, they viewed $\frac{5}{8}$ as the sum of $\frac{1}{2}$ and $\frac{1}{8}$.) The introduction of irrational numbers was so controversial among the ancient Greeks that some historians believe the Pythagoreans murdered one of their own who discovered their existence.

The work of mathematicians often involves making an attempt to solve a problem, finding a refutation to that line of reasoning, and then seeking a new route to solve the problem. These erroneous attempts are in stark contrast to the polished examples that students see in textbooks or worked out by teachers. Devlin (2010) suggested using historical letters between the notable mathematicians Pascal and Fermat can provide students with insights about how uncertainty exists in problem solving, even among great mathematical minds.

Another promising method for teaching the tentativeness of mathematics is the use of rough-draft talk during the act of problem solving. Jansen (2009, Jansen, Cooper, Vascellaro, & Wandless, 2016/2017) discusses how using "rough-draft talk" can provide opportunities for students to express their false starts and uncertainty in their thinking. Jansen's intent of introducing rough-draft talk is to create practices in which all students' thinking is valued and to increase students' participation in mathematical discussions in the classroom. However, when a student engages in rough-draft talk, they also have the opportunity to understand the tentative nature of mathematics. Jansen et al. (2016/2017) identify that one of the underlying principles for rough-draft talk should be the promotion that learning mathematics involves revising one's understanding over time.

4.6 Mathematics Is Creative

An absolutist might argue that mathematics is inherently not creative. If one sees mathematics through the eyes of pre-determined certainty, then creativity in mathematics would be unnecessary. The way that mathematics has traditionally been taught reflects this viewpoint where a teacher describes a procedure or algorithm, and students learn to mimic that procedure. In such a classroom, there is little need for creativity, or what Pair (2017) refers to as "the exploration of ideas."

However, engaging in true problem-solving experiences can help students appreciate the creativity involved in mathematical thinking. For example, a group of teachers was posed the following mathematical task (adapted from Wilburne, 2014) during a professional development workshop: "A fast food restaurant sells chicken tenders in packs of 4 and 7. What is the largest number of tenders that you cannot buy? How do you know this is the largest number that you cannot buy?" This was a task that was unfamiliar to the teachers, and initially the teachers were uncertain whether there would be a largest number that cannot be bought in packs of either 4 or 7. The teachers worked in small groups to solve their problem and document their thinking. One group of teachers (whose work is shown in Fig. 4.1) documented lists of numbers that could be bought and numbers that were impossible to buy using packs of 4 and 7. They realized that four consecutive numbers of 18, 19, 20, and 21 chicken tenders were possible to buy. Given that any of these were possible the next four consecutive numbers – 22, 23, 24, and 25 – could be bought by buying an additional pack of 4 chicken tenders. Since this line of reasoning could be extended, they determined the largest number that could not be bought was 17. A second group (work shown in Fig. 4.2) took a more methodical approach to arrive at the answer

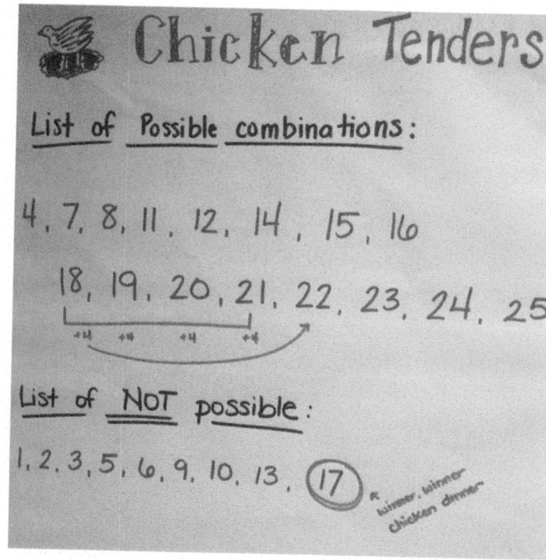

Fig. 4.1 Group 1's Work on the Task

Fig. 4.2 Group 2's Work on the Task

of 17. They listed all the numbers from 1–100 and began by crossing through the multiples of four and multiples of seven to represent that these numbers of chicken tenders could be bought. They continued by crossing through multiples of 11, 15, and 18. After a conversation with the facilitator, they reasoned that since they had crossed off 15 previously, they could also cross of 19 by buying another pack of four. Other numbers were eliminated using similar reasoning. A third group of teachers (work shown in Fig. 4.3) reasoned about the problem by using the expression $4x + 7y$ and created a table of values for x, y, and the total number of tenders.

Fig. 4.3 Group 3's Work on the Task

The type of thinking required to solve this problem stands in stark contrast to the thinking that is typical of traditional school mathematics. The work completed to craft these solutions suggest that the solving required ingenuity. They required creativity and innovation within the minds of the teachers engaged in solving them, and permitted the solvers to take ownership of their creative solution strategies.

The differences in the three groups' methods demonstrate how creativity can be used in mathematical problem solving, but creativity can also play a role in computational tasks. In recent years, the introduction of Number Talks (Parrish, 2014) in mathematics classrooms has provided a way to express students' personal creativity in computation. Although many variations to Number Talks exist, a common routine includes (1) a teacher poses a computational problem; (2) students mentally solve the problem in more than one way; (3) the teacher solicits possible solutions; and (4) students defend their solutions by describing different strategies. These solutions often correspond to a student-created mental strategy, rather than an algorithm. For example, when solving 84–68, one student might say she took 84 minus 60 to get 24, and then subtracted another 8 to get 16. A second student might say he took 84 minus 70 to get 14, and compensated by adding 2 to get 16. A third student might use an adding-up strategy by starting at 68, then adding 2, then adding 10, then adding 4 to arrive at 84. She would recognize that 2 + 10 + 4, or 16, is the difference.

4.7 Future Directions for the Nature of Mathematics Within the Context of STEM Education

In 2017, the National Council of Teachers of Mathematics published the *Compendium for Research in Mathematics Education* (Cai, 2017), an edited volume with 1000+ pages and 38 chapters summarizing the major research completed in mathematics education for the past 10+ years. Although the index is quite detailed, there was no reference to "nature of mathematics," suggesting that the research regarding this topic has not become mainstream among researchers in mathematics education. There is still much to learn about what students and teachers think about the nature of mathematics and how these views impact their work.

STEM provides further opportunities for students to use mathematics in productive ways, but does STEM activity provide opportunities for students to reflect on the nature of mathematics? The answer is inconclusive and is likely context specific. Although STEM has the potential to create rich problems to apply mathematical thinking, there are some mathematics educators with reservations about STEM. They believe that science, technology and engineering may overshadow the mathematical thinking, or that mathematics will be integrated into STEM using mathematics that students have learned previously (e.g., computation, measurement skills). Furthermore, the different disciplines that contribute to STEM may have competing interests. A concern may exist that students will not come to accept mathematics as a way of knowing. However, such assertions can be refuted by emphasizing the definition of STEM as an interdependence of its constituent fields. As such, mathematics should complement and support the other disciplines, just as the other STEM disciplines should complement and support the learning of mathematics. Such hypotheses about the role that mathematics may play in STEM activity provides an important area for further research endeavors. Whether our teaching falls on the side of the absolutist, fallibilists or somewhere in between, STEM's hope for improving mathematics education lies in deliberate opportunities within a STEM context for students to explore and reason about big mathematical ideas. In doing so, they will come to see that their arguments have to be based on assumptions. By examining their assumptions and listening to alternatives, they might come to see mathematics as a creative outlet.

References

Australian Curriculum, Assessment and Reporting Authority. (n.d.). Key ideas. *Australian Curriculum*. https://www.australiancurriculum.edu.au/f-10-curriculum/mathematics/key-ideas/

Boaler, J. (2015). *What's math got to do with it?: How teachers and parents can transform mathematics learning and inspire success*. New York: Penguin.

Cai, J. (Ed.). (2017). *Compendium for research in mathematics education*. Reston, VA: National Council of Teachers of Mathematics.

Davison, D. M., & Mitchell, J. E. (2008). How is mathematics education philosophy reflected in the math wars? *The Mathematics Enthusiast, 5*(1), 143–154.

Devlin, K. (2010). The Pascal-Fermat correspondence: How mathematics is really done. *Mathematics Teacher, 103*, 578–582.

Duval, R. (2006). A cognitive analysis of problems of comprehension in a learning of mathematics. *Educational Studies in Mathematics, 61*(1–2), 103–131.

Ernest, P. (2018). The philosophy of mathematics education: An overview. In P. Ernest (Ed.), *The philosophy of mathematics education today* (pp. 13–37). Hamburg, Germany: Springer.

Healy, L., & Hoyles, C. (2000). A study of proof conceptions in algebra. *Journal for Research in Mathematics Education, 31*, 396–428.

Hersh, R. (1997). *What is mathematics, really?* New York: Oxford University Press.

Jansen, A. (2009). Prospective elementary teachers' motivation to participate in whole-class discussions during mathematics content courses for teachers. *Educational Studies in Mathematics, 71*, 145–160.

Jansen, A., Cooper, B., Vascellaro, S., & Wandless, P. (2016/2017). Rough-draft talk in mathematics classrooms. *Mathematics Teaching in the Middle School, 22*, 304–307.

Jeannotte, D., & Kieran, C. (2017). A conceptual model of mathematical reasoning for school mathematics. *Educational Studies in Mathematics, 96*, 1–16.

Kobiela, M., & Lehrer, R. (2015). The codevelopment of mathematical concepts and the practice of defining. *Journal for Research in Mathematics Education, 46*, 423–454.

Lakatos, I. (1976). *Proofs and refutations: The logic of mathematical discovery*. Cambridge, UK: Cambridge University Press.

Lannin, J., Ellis, A. B., & Elliott, R. (2011). *Developing essential understanding of mathematical reasoning in prekindergarten-grade 8*. Reston, VA: National Council of Teachers of Mathematics.

Ministry of Education, Singapore. (2012). *Mathematics syllabus: Primary one to six*, Curriculum Planning and Development Division, Ministry of Education, Singapore. https://www.moe.gov.sg/docs/default-source/document/education/syllabuses/sciences/files/mathematics_syllabus_primary_1_to_6.pdf

Nathan, M. J., & Petrosino, A. (2003). Expert blind spot among preservice teachers. *American Educational Research Journal, 40*, 905–928.

National Council of Teachers of Mathematics. (2014). *Principles to actions: Ensuring mathematical success for all*. Reston, VA: Author.

National Governors Association Center for Best Practices & Council of Chief State School Officers. (2010). *Common core state standards for mathematics*. Washington, DC: Authors.

Pair, J. D. (2017). *The nature of mathematics: A heuristic inquiry*. Unpublished doctoral dissertation. Middle Tennessee State University, Murfreesboro.

Parrish, S. (2014). *Number talks: Helping children build mental math and computation strategies, Grades K-5*. Sausalito, CA: Math Solutions.

Roach, M., Creager, M., & Eker, A. (2016). Reasoning and sense making in mathematics. In P. Kloosterman, D. Mohr, & C. Walcott (Eds.), *What mathematics do students know and how is that knowledge changing? Evidence from the National Assessment of Educational Progress* (pp. 261–293). Charlotte, NC: Information Age Publishing.

Russell, S. J. (1999). Mathematical reasoning in the elementary grades. In L. V. Stiff (Ed.), *Developing mathematical reasoning in grades K-12, 1999 Yearbook of the National Council of Teachers of Mathematics (NCTM)* (pp. 1–12). Reston, VA: NCTM.

Stylianides, G. J., Stylianides, A. J., & Weber, K. (2017). Research on the teaching and learning of proof: Taking stock and moving forward. In J. Cai (Ed.), *Compendium for research in mathematics education* (pp. 237–266). Reston, VA: NCTM.

Van Hiele, P. M. (1985). The child's thought and geometry. In D. Fuys, D. Geddes, & R. Tischler (Eds.), *English translation of selected writings of Dina van Hiele-Geldof and Pierre M. van Hiele* (pp. 243–252). Brooklyn, NY: Brooklyn College, School of Education. (Original work published 1959).

Weber, K. (2001). Student difficulties in constructing proofs: The need for strategic knowledge. *Educational Studies in Mathematics, 48*(1), 101–119.

White-Fredette, K. (2010). Why not philosophy? Problematizing the philosophy of mathematics in a time of curriculum reform. *The Mathematics Educator, 19*(2), 21–31.

Wilburne, J. (2014, September 15). *What is the largest number you cannot make?* [Blog Post] Retrieved from https://www.nctm.org/Publications/Teaching-Children-Mathematics/Blog/What-is-the-Largest-Number-You-Cannot-Make_/

Rick A. Hudson is an Associate Professor and Chair of Mathematical Sciences at the University of Southern Indiana, and a former secondary teacher. His research interests include the teaching and learning of data analysis and statistics in K-12 schools, mathematics teacher education, and mathematical teaching practices. He is currently co-PI on the NSF-funded grant titled Enhancing Statistics Teacher Education with E-Modules.

Mark A. Creager is an Assistant Professor of Mathematics at the University of Southern Indiana, and a former secondary teacher. His research focuses on preservice and inservice teachers' mathematical knowledge and in particular about how teachers use their mathematical knowledge in classroom settings. Additionally, he investigates students' understandings of proof in geometry and ways of developing an understanding of proof in geometry that is meaningful for students.

Angela Burgess is a Doctoral Student of Science Education at Indiana University and an informal environmental educator. Her research focuses on environmental literacy, environmental education practices and participant outcomes.

Alex Gerber is a Doctoral Candidate in Science Education at Indiana University. He has experience teaching in a variety of informal contexts as well as college courses in the School of Education at Indiana University. His research focuses on informal science education and, more recently, the use of representations in science teaching.

Part II
Critical Questions in Teaching STEM

Chapter 5
Inquiring into Environmental STEM: Striving for an Engaging Inquiry-Based E-STEM Experience for Pre-Service Teachers

Angela Burgess and Gayle A. Buck

5.1 Introduction

Many scholars have described the importance of science, technology, engineering and mathematics (STEM) knowledge and skills for a nation's future economic prosperity and technological competitiveness in global markets (see Blackley & Howell, 2015; Bybee, 2013). Although some fear a potential consequence of this focus on STEM innovations may be global environmental issues, Bybee (2013) argues that the integrated disciplinary approach STEM offers (Kennedy & Odell, 2014) may be useful in addressing global environmental challenges such as climate change, energy production and environmental health. The North American Association for Environmental Education (NAAEE) have proposed an integrated approach to education for sustainable development via Environmental STEM (E-STEM) that aligns with their identified educational best practices including hands-on learning, tangible themes, student interest and fostering achievement and empowerment (Fraser et al., 2013). The NAAEE defines E-STEM as "the integration of environmental education into STEM learning" (NAAEE, 2020).

For students to become E-STEM literate, however, teachers need the content and pedagogical knowledge in order to be able to effectively instruct, assess and design STEM curricula (Shernoff, Sinha, Bressler, & Ginsburg, 2017). This task may not be as simple as it first appears (Akerson et al., 2018). Birney and Cronin (2019) explained, "Beyond the familiar vocabulary of job training, linked learning and twenty-first century skills, STEM teachers must create a combined learning experience that has no precedent in education" (P.1). This lack of precedence means that most teachers have had little direct experience of STEM education (Awad & Barak, 2018) and so may struggle to implement STEM education in their classrooms

A. Burgess (✉) · G. A. Buck
Indiana University, Bloomington, IN, USA
e-mail: ahburges@iu.edu; gabuck@indiana.edu

V. L. Akerson, G. A. Buck (eds.), *Critical Questions in STEM Education*, Contemporary Trends and Issues in Science Education 51,
https://doi.org/10.1007/978-3-030-57646-2_5

(Kelley & Knowles, 2016). Many studies investigate how teacher professional development programs variably equip teachers with the skills they need to successfully implement STEM instruction in their classrooms (e.g. Guzey, Moore, & Harwell, 2016; Slavit, Nelson, & Lesseig, 2016; Stohlmann, Moore, & Roehrig, 2012), while only a few studies have investigated how pre-service teachers' experiences of STEM impact their conceptualisations and instruction of STEM (e.g. Adams, Miller, Saul, & Pegg, 2014; Awad & Barak, 2018; Berry, McLaughlin, & Cooper, 2018); fewer still have investigated this in regard to E-STEM.

The aim of this study was to examine how pre-service teachers experience an E- STEM intervention in a science content course for elementary education majors. The context of a content course was selected due to the fact that the content knowledge of K-12 teachers needs to be increased in addition to their pedagogical knowledge (Honey, Pearson, & Schweingruber, 2014). There is a dearth of research into pre-service teachers STEM experiences in these contexts, despite content knowledge being recognised as an integral aspect of an individual's pedagogical content knowledge. The overall research question we addressed was; In what ways does the inclusion of an E-STEM intervention in an elementary education science content course impact pre-service teachers? The sub questions guiding this action research were, 1) How does our E-STEM intervention influence pre-service teachers understanding of STEM? 2) To what extent does our E-STEM intervention impact pre-service teachers' notion of environmental issues?

5.2 Theoretical and Empirical Background

What STEM actually means is up for debate (Shernoff et al., 2017). Individuals perceive STEM differently and this perception is often influenced by one's role within the education system (Bybee, 2013), discipline area, or how they use STEM in their everyday lives (Breiner, Harkness, Johnson, & Koehler, 2012). Perspectives range from STEM as it's individual domain constituent (e.g. STEM is Science) to STEM as a meta-discipline that fully integrates all four constituent disciplines (Bybee, 2013; Kennedy & Odell, 2014). Shernoff et al. (2017) argue that integrated STEM education definitions should be considered within a conceptual framework which considers the interactions between the goals, outcomes and intentions of integrated STEM education. Kelley and Knowles (2016) propose integrated STEM education should consist of STEM practices from each constituent discipline including scientific inquiry, mathematical thinking, technological literacy and engineering design in order to foster situated STEM learning.

The intervention in this study was based upon our working definition of STEM as meaningful interdependence among all disciplines of STEM. In other words, includes all individual disciplines of STEM (science, technology, engineering, and mathematics) in a way that is meaningful and showcases the interdependence of the fields. In an education setting this means that any STEM educational experience must be situated within an authentic context that makes the presence and *meaningful* interdependence of each discipline explicit. This definition does not necessarily exclude integrated STEM education approaches if all four disciplines are expressed. We would argue that combinations of only two to three disciplines constitutes interdisciplinary education, but not STEM education. However, as STEM conceptualisations are so various, we do not discount our participants' conceptions of STEM as incorrect if they do not align with our own. Although this interdependent view of STEM may seem to best reflect authentic practices (Ring, Dare, Crotty, & Roehrig, 2017), many researchers have discussed the difficulties teachers have implementing this type of STEM education in their classrooms (e.g. English, 2016; Rinke, Gladstone-Brown, Kinlaw, & Cappiello, 2016). Nadelson et al. (2013) specifically cite lack of teacher preparedness as an obstacle to providing genuinely integrated STEM education. In this study, we investigated how pre-service teacher content courses may contribute to preparing teachers to teach interdependent STEM.

Like STEM, environmental education is ambiguous (Stevenson, Brody, Dillon, & Wals, 2013) and conceptualised differently throughout the literature. We used environmental education as an umbrella term that we define as Vare and Scott's (2007) interrelated idea about education *for* sustainable development, and education *as* sustainable development. This model of environmental education allows us to consider the existing environmental issues as socially and politically induced as well as accepting that we can not fully appreciate the requirements of future societies, and so must embrace open-ended learning. The second point is similar to the skills acquisition and conceptual framework notions of STEM. Tilbury's UNESCO report (Tilbury, 2011) suggested some key learning processes of environmental education include collaboration, whole system engagement, innovation and active, and participatory learning, processes we argue are consistent with the twenty-first century skills for which STEM education has been credited with influencing (e.g. Bybee, 2013; Ostler, 2012). Researchers in environmental education have also discussed the value of using integrated, interdisciplinary approaches (e.g., Howlett, Ferreira, & Blomfield, 2016; Jones, Selby, & Sterling, 2010), a key feature of STEM education. The similarities in the conceptual frameworks of STEM education and Vare and Scott's ideas about education for/as sustainable development have led some technology and design education scholars to suggest that there is a blurring of the boundaries between the meta-disciplines (Pitt, 2009).

As in STEM education, a pervasive issue with environmental education implementation in schools is the teacher's perceived lack of knowledge of issues and therefore diminished self-efficacy to teach about environmental sustainability. Higher education is viewed as an important site at which graduates can experience environmental education that may impact their future professional lives (Holdsworth & Thomas, 2016). Provision of higher education experiences that have a content focus on sustainability can result in significant changes in pre-service teachers' environmental perceptions and values, as well as improve their agency and motivation to include environmental sustainability topics in their classroom (Merritt, Hale, & Archambault, 2019). As educators are often viewed as positive role models, it is assumed that they have a substantial influence on community practices via their contact with students, parents and other community members (Anderson, Datta, Dyck, Kayira, & McVittie, 2016). STEM and environmental literacy are placed as critical aspects in a dynamic future-focused society. Given the similar attributes a STEM and environmentally literate person should possess, as well as the lack of preparation pre-service teachers receive in these areas, it seems pertinent to explore how using environmental sustainability as a focus of STEM instruction in a teacher preparation, higher education context may impact teacher conceptions. Holdsworth and Thomas (2016) state that, "A lack of reflection on one's practice will fail to transform practice into praxis, reinforcing the current reductionist, individual approach to education seen today" (p. 1077). In this study we aim to identify if and how pre-service teacher (PST) participation in an E-STEM higher education experience influences their conceptions of STEM and sustainability.

5.3 Methodology

This study was conducted as an action research (Lewin, 1948). Action research is the cyclical approach of making change, analysing that change for effectiveness and making further improvements to the action (Eilks, 2018). Action research connects research with practice by having classroom practitioners become classroom researchers. This type of research intends to improve classroom practices as well as contributing to the practitioner's professional development (Feldman, 1996). Yet, it should also be viewed as a medium through which we can validate strategies for educational innovation (Eilks & Ralle, 2002). The critical reflective practices that are invoked in action research is advocated for in the development of teachers of sustainability and sustainable development (Wals & Jickling, 2002). The action research undertaken in this study was one of a teacher-centered approach (Grundy, 1982) where the practitioners were responsible for deciding the research interest, classroom action, data collection and interpretation and deciding the implications for the action. In this study, both authors co-designed, utilised and analysed the intervention in their own classrooms alongside a third practitioner who implemented the intervention in two additional classrooms. The primary purpose of this action research was to address the problem of insufficient conceptual understanding teachers hold of both STEM (as determined via a review of the literature) and to investigate how to implement and design STEM interventions that are rooted in environmental issues by introducing an E-STEM intervention. Considering this, we view this study as exploratory and will use it to inform future iterations of the intervention. The choice of an action research approach centres learning from experience (Dewey, 1986), in this case the experiences of the practitioner/researcher, as well as the pre-service teachers, while recognising the importance of practitioner reflexivity and reflection in the reformation of educational practices.

5.4 Participants

Our pool of participants included undergraduate students enrolled in a general education science content course in a large Midwestern university in the USA. In total, there were 81 participants from a cohort of 93 students enrolled in the 4 classes where the intervention was implemented giving a participation rate of 87%. Of the 81 participants, 89% were female (n = 72), and 11% were male (n = 9). The students identified as either freshmen (n = 43), sophomore (n = 32), juniors (n = 5) or seniors (n = 1).

5.5 Context

The intervention took place during the Fall semester within a general education science content course. This is a required course for elementary education majors and an optional course for non-elementary majors. Therefore, the majority (87%) of the students enrolled in this course during this study were education majors. The course is rooted in environmental science and scientific inquiry content. The course is designed using socio-constructivist principles of learning and is split into three broad sections. The first section explores nature of science and principles of scientific inquiry; the second section introduces the use of scientific explanations in scientific inquiry using several environment focused inquiry scenarios and labs; and, in the final section, students engage in a free-choice scientific inquiry and produce a full scientific report and presentation based on their independent scientific inquiry. The E-STEM intervention took place during the second section of the course.

5.6 Action Description

The E-STEM intervention described in this study was based upon the foundations of social constructivist theories (Vygotsky, 1978). The key principles of constructivist theories of knowing and learning are that learning is an active process and each learner enters that process with a schema or mental model that is based on their prior knowledge and experiences (Bruner, 1966). Learning is the reconfiguration of these schemata into something more similar to that of an expert. Vygotsky's (1978) social constructivist theories of knowledge is based upon these premises but introduces the notion that the collaborative interaction between individuals develops more comprehensive schemata in those collaborators. The aim of this intervention was to provide a collaborative learning environment and promote reflection of experience (Dewey, 1986). By working collaboratively, it was theorised that the pre-service teachers would develop more comprehensive schemata than would otherwise be constructed. As the aim of this intervention was to promote conceptual change, a collaborative project-based-learning instructional approach was used (Kelley & Knowles, 2016).

The task, as established in class 2 (see Table 5.1), was: "Design and build a functional 1:10 scale model of a solar water heater that will allow you to provide 10% of your daily hot water requirement. The water should be heated to at least 45 Degrees Celsius.". The intervention took place over four class periods of one hour and fifty-five minutes. Homework assignments related to the topic were assigned prior to the implementation of the classroom interventions and the project culminated in a report writing assignment. For details of the intervention, see Table 5.1.

Table 5.1 Description of E-STEM intervention

	Class 1	**Class 2**	**Class 3**	**Class 4**
Activity	Introduction to heat transfer	Solar energy and project task introduction	Initial design plan, build and test	Design modification and testing
Purpose	Provide students with needed scientific concepts for design task	Establish connections between concepts, environment, sustainability, STEM and the task.	Use collaborative design process Consider material property appropriateness. Apply scientific concepts to design Evidence-based design	Use data to compare efficacy of design iterations Identifying consequences of design modifications
Target concepts	Radiation, conduction and convection	STEM	Applying science concepts in design	Applying science concepts in design
		Sustainability	Evidenced-based design	Evidence based design
		Energy	Use of technological tools to provide evidence	Use of technological tools to provide evidence. Design process is iterative
	Homework 1	Homework 2	Homework 3	Homework 4
Activity	Hot water use analysis	Solar water heater designs	Analyse data and modifications	Write a reflective technical report based on rubric
Purpose	Provide real world data from which to base model design Highlight consumptive behaviours	Consider current solar technologies	Use data to determine the efficacy of the initial design Consider design modifications to improve efficacy	Assessment of E-STEM conceptualisations
Target	Energy consumption Calculating scales	Technological progress is built upon prior technologies	Design process is iterative	Integrated E-STEM concepts

5.7 Data Collection and Analysis Procedures

Multiple sources of data were used to allow for triangulation and a more sophisticated insight into student experiences as a result of the action. The data generated was both qualitative and quantitative which lets us not only reveal general patterns of change across the participant population, but also to understand the nuances and rich points of the participant experiences.

Pre and Post Survey To understand broad changes in students' conceptions and attitudes towards STEM as a result of the action, a pre-survey was completed by participants one week prior to the commencement of the intervention (n = 81) and a post survey was completed one week after the completion of the intervention (n = 81). The survey was adapted from Summers and Abd-El-Khalick's (2018) BRAINS (Behaviours, Related Attitudes, and Intentions towards Science) survey to measure changes in five constructs/domains related to STEM conceptualisations. These constructs/domains were intention, attitude toward the behaviour, behavioural beliefs, control (beliefs and perceived behavioural control) and, normative (beliefs and subjective norms) deriving from the theory of planned behaviour (Ajzen, 1985) (See Table 5.2). The survey contains 30 statement items scored on 1–5 Likert scale ranging from 1-Strongly disagree to 5-strongly agree. The survey is provided in Appendix A. The survey was administered online using Google Forms. As well as the 30 items adapted from the BRAINS survey, participants were asked to briefly describe what the term STEM meant to them, and, in the pre-survey

Table 5.2 The constructs measured using the adapted BRAINS survey

Construct	Related domain/construct	Sub-Domain/Sub-construct
Intention	Intention to pursue or interest in pursuing STEM	Additional or future studies in STEM A career in STEM
Attitude toward the behaviour	Attitude toward different facets of STEM as relates to respondent's life	Attitude toward STEM Attitude toward STEM as leisure
Behavioural beliefs	Beliefs about the consequences associated with engagement with STEM and beliefs about the benefits associated with STEM	Beliefs about the consequences associated with STEM learning Beliefs about the relevance and utility of STEM at the societal and personal level
Control- beliefs and perceived behavioural control	Perceived self-efficacy and personal agency toward STEM learning	Perceived ability towards learning STEM Perceived efficacy of effort toward learning STEM
Normative- beliefs and subjective norms	Perceived approval or disapproval toward engagement with STEM	Perceived approval or disapproval by family and friends

Modified from Summers and Abd-El-Khalick (2018)

only, they were asked to describe any prior experience they had with STEM. Demographic information was also collected via the online survey.

The 30 items of the adapted BRAINS survey were categorised under their corresponding psychological construct as prescribed by Summers and Abd-El-Khalick (2018). Two items were reverse coded and then an independent t-test was performed using IBM SPSS to look for changes between pre-survey and post survey responses. Participant written descriptions of STEM were coded using emergent thematic coding. Code frequencies were then input into IBM SPSS and an independent t-test was performed in order to highlight differences in the frequencies of particular codes between the pre and post intervention surveys.

Student Work Students produced several pieces of work throughout the intervention including: 1) scientific explanations of heat transfer observations; 2) design diagrams for both the initial and modified scale model design by each group of students; 3) worksheets where each group outlined the scientific and practical justifications of their material choices; and, 4) individually produced final reports based on the rubric in Appendix B. The initial stage of analysis of student work involved a first pass over the work while noting areas of salience and interest in memos (Glaser, 1978). Once this stage was completed, primary descriptive codes (allowing for simultaneous coding) were determined, and the student work was coded using NVivo qualitative analysis software. These codes were then analysed for patterns and categorised based upon emergent themes.

Audio Recordings As collaboration and interaction between participants are key to the socio-constructivist theories that informed this intervention, it is appropriate to record the naturally occurring conversations that occur as a result. Each group of students were recorded using an audio recorder placed at their workspace. Student groups typically consisted of 3 to 4 students. The interactions of 24 groups were recorded in all four class periods, over the four sections, resulting in a total of 182.4 hours of recorded classroom interaction. From each section, audio recordings of 2 student groups were selected and transcribed for analysis (totalling 60.8 hours). The selection of groups was based upon several factors. The first factor was that each member of the group had given consent for their recording to be analysed; the second factor was that each member of the group was present in all four class periods of the intervention; and the third selection criteria was that the quality of audio recording was high enough to allow for accurate transcription.

The audio recordings were played back, memos were produced that recorded initial interpretations and patterns identified in the data. Recordings were revised several times and relevant sections of conversational interaction were transcribed verbatim. For an interaction to be identified as relevant the conversation should be directed toward the materials, design, concepts or generally focused on the task. The transcripts produced were then coded descriptively and themes were developed from patterns across the codes.

5.8 Findings

The findings are presented in relation to the two guiding research questions. Any additional interesting patterns that were observed are also pointed out.

5.8.1 *How Does Our E-STEM Intervention Influence Pre-Service Teachers' Understanding of STEM?*

Independent samples t-tests of the BRAINS survey revealed that there was statistically significant positive change in three of the five constructs measured in the post-intervention survey compared to the pre-intervention survey. The attitude component showed a significant positive change at a 5% level from pre-intervention ($M = 7.9$, $SD = 5.2$) to post intervention ($M = 10.5$, $SD = 3.7$), $t(160) = 3.527$, $p < .001$. The 95% confidence interval for the difference in means ranged from −3.9 to −1.1. The effect size was medium according to Cohen's guidelines, g = 0.576. The behaviour component also exhibited a significant positive increase, t(160) = 2.327, p < 0.05, between pre intervention (M = 16.6, SD = 8.5) and post intervention surveys (M = 19.4, SD = 7.0). The 95% confidence interval for the difference in means ranged from −5.2 to −0.4 with a medium effect size, g = 0.360. Finally, the control component exhibited a significant change from the preintervention survey (M = 8.9, SD = 5.3) to the postintervention survey (M = 11.5, SD = 4.3) t(160) = 3.425, p < 0.05. The 95% confidence interval ranges from −4.1 to −1.1 with a medium to large effect size, g = 0.539. Although the normative and intentional components of the survey did change slightly between the pre and the post survey, neither changes reached a level of significance (Table 5.3).

These findings corresponded with the statistical analysis of the STEM definitions participants produced before and after the intervention. Each response to the question "Briefly describe what does STEM mean to you?" from both the pre and post intervention survey was coded using emergent codes. A total of 16 codes were identified (Table 5.4). Individual statements could be coded with multiple codes. Each code was then compared to investigate any differences in code frequency between pre and post intervention statements using independent samples t-test. A significant

Table 5.3 Results from independent t-test between pre and post construct conceptualisations

	Pre		Post		95% CI				
Variable	*M*	*SD*	*M*	*SD*	*t*(160)	*p*	*LL*	*UL*	*Hedge's g*
Attitude	7.94	5.20	10.48	3.74	−3.57	<.001	−3.95	−1.14	0.56
Intention	7.88	3.99	8.72	3.79	−1.37	.17	−2.05	0.37	0.22
Behaviour	16.59	8.45	19.43	7.01	−2.32	.02	−5.25	−0.43	0.37
Control	8.91	5.33	11.52	4.30	−3.43	.001	−4.11	−1.10	0.54
Normative	4.54	2.74	5.17	2.22	−0.85	.39	−1.10	0.44	0.25

Table 5.4 Results of an independent t-test between coded pre and post intervention definitions of STEM

	Pre		Post		95% CI				
Code	*M*	*SD*	*M*	*SD*	*t*(164)	*p*	*LL*	*UL*	*Hedge's g*
Important	0.02	0.15	0.12	0.33	−2.43	.016	−0.17	−.018	0.38
Research	0.01	0.11	0.02	0.15	−0.58	0.56	−0.05	0.03	0.08
Nothing	0.04	0.19	.048	0.22	−0.38	0.70	−0.07	0.05	0.04
Integrated	0.06	0.24	0.22	0.41	−2.98	.003	−0.26	−0.05	0.46
Separate	0.02	0.15	0.00	0.00	1.42	0.16	−.009	0.58	NA
Science	0.43	0.50	0.50	0.50	−0.75	0.44	−0.25	0.04	0.14
Maths	0.28	0.45	0.39	0.49	−1.49	0.14	−0.25	0.36	0.23
Technology	0.22	0.41	0.41	0.49	−2.72	.007	−0.33	−.053	0.43
Engineering	0.19	0.40	0.40	0.49	−2.95	.004	−0.34	−.068	0.46
Skills	0.06	0.24	0.14	0.35	−1.80	0.07	−0.18	.008	0.27
I don't know	0.14	0.35	0.02	0.15	2.84	.005	0.12	0.04	0.44
Acronym	0.36	0.48	0.01	0.11	6.42	.000	0.24	0.46	0.99
Workforce	0.36	0.19	0.36	0.19	0	1.00	−0.58	0.58	0
No resp.	0.12	0.33	0.16	0.37	−.671	.503	−0.14	0.07	0.13
Real-life	0.05	0.22	0.23	0.42	−4.47	.001	−0.28	−0.08	0.54
Education	0.16	0.37	0.06	0.24	2.00	0.46	.002	0.19	0.3

difference between frequencies was detected for 7 codes. The frequency at which STEM was described as important was significantly higher t(164) = 2.426, p < 0.05 in the post intervention survey (M = 0.1205 SD = 0.32750), than in the preintervention survey (M = 0.0241, SD = 0.15428). The 95% confidence interval ranged from −0.175 to −0.018 with a small to medium effect size of g = 0.377. The code that indicated if a definition of STEM included an integrated view of STEM also showed significant changes t(164) = −2.98 p < 0.01, between pre (M = 0.060, SD = 0.239) and post engagement (M = 0.217, SD = 0.415)in the E-STEM activity with a small-medium effect size of g = 0.463. There was a significant change in the inclusion of specific disciplines with STEM definitions. Technology was cited as significantly more frequently t(164) = −2.721, p < 0.01 in the post intervention definitions (M = 0.41, SD = 0.49) than in the pre- intervention definitions (M = 0.217, SD = 0.41462) with a small to medium effect size of 0.425. Similarly, Engineering was also cited significantly more frequently t(164) = −2.951, p < 0.01 post intervention (M = 0.40, SD = 0.49, preintervention M = 0.193, SD = 0.40) with a small to medium effect size of 0.46. The frequency at which students reported that they did not know a definition of STEM decreased significantly t(164) = 2.844, p < 0.01 post intervention (M = 0.024, SD = 0.154 Preintervention M = 0.145, SD = 0.354) with a small to medium effect size of 0.44. The frequency of describing STEM only in terms of the acronym constituents significantly decreased, t(164) = 6.422, p < 0.01, between pre (M = 0.361 SD = 0.483) and post (M = 0.012, SD = 0.11)intervention with a large effect size of 0.99. Participants mentioned real-life application in their definitions of STEM significantly more frequently, t(164) = −3.47, p < 0.01, after the intervention (M = 0.23, SD = 0.423) than before (M = 0.0482, SD = 0.22) with a medium effect size of 0.54. Finally, the frequency of definitions that included a reference to STEM as an educational approach significantly decreased, $t(164) = 2.009$, $p < 0.05$, post intervention ($M = 0.0602$ $SD = 0.23938$, Preintervention $M = 0.157$, $SD = 0.36566$) with small effect size of $g = 0.3$.

The production of a final paper provided students with an opportunity to reflect on how they used STEM in order to complete the set task of building a functional 1:10 scale solar water heater model. This reflection was prompted by the rubric requirements (Appendix B). Students reflected on STEM in their papers in a variety of ways. Some reflected upon their use of skills from the different disciplines while describing their designing and building process and so referred to STEM throughout their reports. One example follows.

Excerpt 1 [description of calculation] *In total, we needed to heat 729 mL of water. This falls into the 'mathematics' sector of STEM*

Excerpt 2 *With the sectors 'science' and 'technology' from our knowledge of STEM, we realized that using aluminum foil would be most successful for creating a convection-based heat source.*

While others dedicated separated sections of their paper to describing how they utilised STEM. For example, one student noted:

> *STEM was used within our design. Science was used when we were thinking about the use of heat transfer concepts (conduction, convection, and radiation), and when collecting evidence based on effectiveness of our design. Technology was used when we were designing and adjusting our water heater, to create a product with a human use, and collect evidence based on the effectiveness of our design. Engineering was used to collect evidence based on the effectiveness of our design, when using concepts of heat transfer, and designing and adjusting our water heater. Lastly, mathematics was used when collecting evidence based on the effectiveness of our design and scaling our daily average down to a 1:10 scale.*

Most students separated the disciplines out when asked to describe how they had utilised STEM in the activity as in examples 1 and 2. Occasionally students would consider the thinking processes and skills as well as the application of the separate disciplines. This suggests the construction of a conceptual framework model (Kelley & Knowles, 2016) of STEM understanding. For example, one student reflected:

> *The last big component of STEM learning is the four C's: creativity, collaboration, critical thinking, and communication. The solar water heater needed all four of these components; first, students had to be creative to come up with a design that would work. Second, the students had to collaborate and communicate with their lab group to successfully make alterations in the design. Students had to critical think to adjust in their solar water heater designs.*

Despite the emphasis on, and the explicit instruction of STEM used in the intervention, the audio recordings of student interaction exhibited no explicit mention of or reflection on STEM or STEM processes. Typically, student conversation centred around the particular discipline they were working in at that time. Occasionally it was possible to determine how students were using both science and engineering concepts, particularly when they were discussing the properties of materials and the use of technologies.

Example 1

Jordan *I don't get the point of the pump.*

Sofie *No me neither.*

Jordan *It just moves the water so I don't know why we need it.*

Aliyah *Isn't that like... convection though.*

Jordan *No.... I think it is literally like... moving it in the tubes.*

Crystal *Like... it is just moving the water.*

Jordan *Like... it is not like... moving it fast enough to... heat it. It's just* like...moving *it through.*

Crystal *So we don't want to use it?*

Jordan *I don't think we need to.*

Example 2

Jess *The tin foil...I wish we had the results from our experiments. [Travis] do you have the results in your notes still?*

Travis *Yeah.*

Jess *Can I see it please?*

Travis *Yes of course.*

Jess *The one where we wrapped the cup in different things.*

Amy *But...were we really that accurate with this?*

Jess *Yes...yeah...thank you... the tin foil got hot*

Amy *Yeah I feel like the tin foil would be better than the foam probably.*

Jess *But what if we wrapped it in tinfoil and then foam and then put the foil on top of this?*

Travis [Jess] *wow...that's an idea.*

Jess *Thank you.*

Overall, the evidence suggests that there is a shift from a siloed notion of STEM as the constituent disciplines toward a more integrated understanding of STEM in terms of both the underlying content and the skills and processes requirements.

5.8.2 *To What Extent Does Our E-STEM Intervention Impact Pre-Service Teachers' Notion of Environmental Issues?*

The BRAINS survey has 30 individual statements to which respondents indicate their agreement using a 5-point Likert scale. One statement is directly related to STEM and the environment. In the pre-intervention survey 62.5% of respondents indicated that they either agreed or strongly agreed with the statement "Knowledge of STEM helps me protect the environment". In the post-intervention test, 75.3% indicated that they agreed or strongly agreed with the same statement. No single respondent indicated they strongly disagreed with the statement in either the pre or post intervention survey, the respondents who indicated that they did not know fell from 36.3% in the pre-intervention survey to 22.2% in the post- intervention survey.

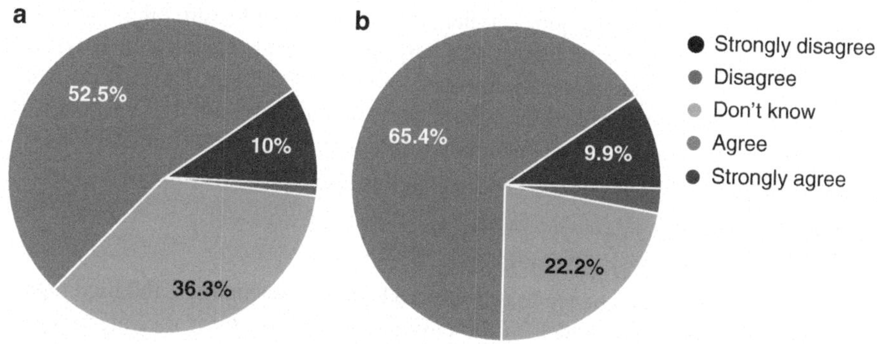

Fig. 5.1 The responses to the statement "Knowledge of STEM helps me protect the environment". (**a**) Pre intervention responses in percentage (**b**) Post intervention responses in percentage

The percentage of respondents who disagreed with the statement rose from 1.2% to 2.5% (Fig. 5.1.). Upon statistical analysis using an independent t-test it was observed that, although there is an increase in the percentage of respondents that agree to some extent with the statement, this increase is not statistically significant at an alpha of 0.95 ($p = 0.07$).

Audio recordings of student interaction suggested that particular aspects of environmental issues were more evident at particular times during the intervention. An area that particularly increased student communication about environmental issues was when students were calculating their individual hot water use. Students often expressed surprise, or pride in their hot water use calculations and related this to an "environmentally friendly" identity.

Ben *Ok Sam, yours is 36.78*
Sam *Is that a lot?*
Ben *No that is less than Laura*
Kasey *No we have like a really small amount*
Sam *I am so environmentally friendly guys*
Laura *I had a lot because I run a dishwasher and I do laundry*
Ben *Same though…*
Laura *There are 4 of us. I am only saying...I mean I don't do other people's laundry... but like*
Sam *I do not… I just did like 3 loads of laundry last night because I washed my towels too*
Laura *So added all together*
Kasey *One hundred and…*
Sam *Wait what was mine? 46 point..*
Laura *So add it all together it is 321.99…divide that by 4*
Ben *yeah 80.5 so we are doing 0.805 litres of water…that's not bad*
Sam *You're welcome (laughs)*

In this example, students connect less hot water use with being more environmentally friendly and those with a higher hot water use attempt to justify their calculations by discussing the consumptive activities in which they take part. There seems to be a general concern about the volume of hot water the individuals use and they spend some time comparing their consumption. Ben states that the small volume of water the group requires to heat is "not bad", while Sam takes credit for reducing their need after establishing that her usage is small. This is a fairly typical interaction between students during this part of the intervention. The concern with reducing consumption does however not seem to extend to material use. There was no mention of economic use of materials throughout the recorded interactions, in fact consumption of material without justification was often evident.

Travis *Why am I painting both sides black?*
Amy *I don't know ...like I have no justification behind it*

As well as offering a chance to reflect on STEM understandings, the final report also allowed students to reflect on the environmental aspects of the intervention. This is where more clear connections were made between the environmental issue and the activity the students took part in. Most students discussed the use of solar energy as a source of renewable energy and connected that to reduced carbon dioxide emissions. One student noted:

By 2100, our energy consumption is expected to triple, leading to an enormous spike in the carbon dioxide emissions every year []. Carbon dioxide is a greenhouse gas, so with this spike in carbon dioxide emission the gases will trap more heat, causing a rise in global temperatures, which can lead to a multitude of problems for our environment. One way we can help cut this energy consumption down is by using sustainable energy, such as solar energy.

Other students reflected on how the specific task in the intervention led to a fuller understanding of the financial and environmental impacts of consumptive behaviour, highlighting student's values relating the development of renewable energy sources.

Creating a water heater made me more aware of how I should not be wasting water or taking it for granted because it takes a long process to heat the water. I also became more consciousness of how much water I was using after this project because since the process is not natural, there is a heavy cost and I learned I have to pay for the amount of water I use.

As indicated in the audio recordings, the part of the activity where students calculated their volume of hot water consumption appeared to be a salient part of the activity.

*Looking at myself, for instance, I found that I use at least 26 liters of water when I shower. Looking at that number, I realized how careless I have been about my water use. This project helped me to see just how difficult it is to heat that much water. Looking at the carbon footprint that human beings leave, using solar panels is more efficient and the energy is free. One of the benefits of solar heating, as quoted, states "Solar hot water is a green, renewable heating system and can reduce your carbon dioxide emissions".*Other students reflected on this part of the activity as well as their membership of a global community.

We found in class that alone, I use 4,319.65 Kilowatts of energy per year solely on heating my own water. I am only one person out of the nearly 8 billion on Earth that are heating their own water as well, which highlights the need for renewable energy sources to heat all of this water.

However, the reflection students engaged in during their report writing also unearthed some naïve conceptions about climate change that students continued to hold after the intervention.

> *As more green-house gasses are damaging our atmosphere more solar rays are reaching Earth's surface. While this is the main contributor to rising global temperatures, it also means there is more potential energy to utilize.*

In fact, naïve conceptions were evident throughout the data and were concerned not only with environmental issues, but also concerning the fundamental scientific concepts referred to in the intervention. For example, students expressed such naïve conceptions as insulation "holding in the cold" and the active movement of water via a pump as constituting convection. A large proportion of students referred to the notion of heat or light "attraction" when describing their justification for painting parts of their design black or the use of aluminium foil in their designs.

5.9 Discussion

The evidence presented indicates that pre-service teachers' (PST) experiences of E-STEM interventions in science content courses does positively impact their conceptualisations of STEM. The changes observed in the definitions the participants constructed before and after the intervention are similar to those reported as a result of a 3-week professional development course for in-service teachers specifically targeting integrated STEM instruction (Ring et al., 2017). Like Ring et al.'s (2017) study, one consequence of this intervention was an increase in participant definitions of STEM that included notions of integration. Another similarity is that engineering featured more prominently in participant's post-intervention definitions as did technology. Simplistic definitions of STEM as an acronym and participants indicating that they did not know how to define STEM reduced significantly. This indicates that a content-based course could significantly contribute to developing PST's conceptualisations of STEM in similar ways to methods-based professional development. Shernoff et al. (2017) found that one of the most cited challenges that teachers face when trying to implement integrated STEM in their classrooms is their lack of understanding of what integrated STEM means. The researchers also described how administrators valued teachers with integrated STEM conceptualisations as a "scarce commodity" (p.9) who are essential in driving STEM education in schools (Shernoff et al., 2017). Our study suggests that the introduction of integrated STEM interventions in the science content classroom may be a fruitful context in which to begin confronting integrated STEM conceptions in future elementary school teachers.

Many teachers feel unprepared to teach STEM (Shernoff et al., 2017) and feel uncomfortable being asked to teach a subject for which they have not been trained (Stohlmann et al., 2012). When examining changes in PST's attitudes towards

STEM as a result of the intervention, there was significant positive change in the behavioural beliefs, attitude to behaviours and beliefs about behavioural control. This suggests that not only are they more informed about the ways in which STEM contributes to their everyday lives, but they have an increased understanding of the beneficial consequences of engaging in STEM and increased self-efficacy and agency in their learning of STEM. Inservice teachers have suggested that more STEM experiences in teacher education may help them feel more prepared to teach STEM (Shernoff et al., 2017) and the results form this survey suggest this intervention does impact PSTs beliefs about their ability to learn about, and, their attitudes toward the utility of, STEM.

Many teachers cite lack of content knowledge as a barrier to implementing integrated STEM instruction in their classrooms (Shernoff et al., 2017). We hoped that by engaging PSTs in integrated STEM activities they would increase their content knowledge and therefore begin to address this issue. However, the findings suggest that although this intervention may have led to improved conceptualisations of integrated STEM, key science, mathematics, technology and engineering content concepts were either naively conceived, poorly communicated by PSTs and/or teacher educators, or the transfer from learning to application context was simply too far (Barnett & Ceci, 2002). Further intervention adjustment and study is required in order to understand this observation, but this iteration of the intervention did little to increase individual discipline specific content knowledge.

There were several aspects of the E-STEM intervention that suggest that PSTs were becoming aware of issues related to the environment and sustainability. The intervention unearthed some of the everyday cultural patterns of a consumptive society (Anderson et al., 2016) and PSTs spent time reflecting on the consequences of that consumption. One viewpoint among many is that environmental education should challenge norms and work to shape people's values in light of global environmental issues (Bell, 2004; Kopnina, 2012). There is evidence to suggest that norms were being confronted, if not challenged, as a result of this intervention. A rich area where consumptive norms were troubled was during the hot water use calculations. People with low consumption were supported, and people with higher consumption spent time justifying their hot water use. Within the student work, PSTs reported reflecting upon their consumptive behaviours, regarding their actions as "careless". These types of shifts in the sustainability related behaviours acceptable to a community of practice (Lave & Wenger, 1991) have been recorded in the literature (Anderson et al., 2016) and potentially transformative experiences such as those provided within this intervention are desired outcome of environmental education in higher education (Cranton, 2006; Howlett et al., 2016). Although an aim of this intervention was to shape PST's environmental values, we endeavoured not to portray a philosophy of "environmental idealism", but to provide them with opportunities to encounter and reflect on their behaviours and experiences within their own environmental value system. That is, that we attempted to promote a "sustainability frame of mind" (Bonnett, 2002). The diversity of PST's attitudes toward sustainability was evident in the reflective section of their final reports that typified both biocentric and anthropocentric environmental value systems.

It is clear that the reflective technical report, the final assessment piece of the intervention, is where the PST's environmental understandings were most evident. Rodgers (2002) interprets Dewey's notion of reflection as making meaning of experiences, to "formulate the relationships and continuities among the elements of an experience, between that experience and other experiences, between that experience and the knowledge one carries, and between that experience and the knowledge produced by thinkers other than oneself." The technical report section of the intervention gave the PSTs the opportunity to reflect upon and make sense of their experiences in relation to their own environmental values and in some cases impacted their beliefs about their sustainable behaviours. The depth of reflection and therefore meaning making and exploration of beliefs was by no means uniform throughout the PSTs and it is important to note that despite apparent successes at increasing awareness, reflection and critical examination of one's behaviours, this may not lead to sustainability action (Festinger, 1957). The next iteration of this action should include more explicit guidance on reflection in the hope of encouraging more PSTs to engage more fully in reflective activities. Another finding related to PST's understanding of environmental issues is that there was also no statistically significant difference in the percentage of students who agreed with the statement that "knowledge of STEM helps me protect the environment". It is unclear if they already held this view prior to the intervention and so the intervention just confirmed their ideas, or if the intervention did do little to change this perception. There is also the possibility that students do not see the connection between the apparent increased awareness of environmental issues, illustrated by the evidence, and STEM. Further work would need to be carried out to reveal the nuances of this relationship.

5.10 Conclusion

This study used an action research approach to explore the provision of an E-STEM higher education experience for PSTs for the purpose of enhancing that experience. Our findings reveal that E-STEM interventions, such as the one under study here, are promising strategies to target these conceptions for improvement. We found that engagement in the intervention allowed PSTs to confront and reflect upon their ideas about STEM which led to improved conceptions of integrated STEM. The study also revealed that, by situating the STEM experience within an environmental context, PSTs reflected on their environmental practices and exhibited positive shifts in attitudes towards consumptive behaviours.

Our findings also revealed areas in which our intervention should be modified. First, the experience needs to be revised in a manner that assures a greater level of content understanding. As mentioned, we found the persistence of naïve understandings in both the scientific and mathematical concepts we hoped would have been used to drive the design process. In future iterations of this intervention we recommend that, prior to designing their own models, PSTs should observe an existing model and identify materials and features designed to convert electromagnetic energy to thermal energy and identify the heat transfer concepts at play. They should also use this model to convert scales to enable them to conceptualise the function and process of using scale to produce models. We predict that the inclusion of this activity into the intervention, we would be able to provide students with more concrete representations of both scale and heat transfer.

Second, the action needs to be revised in a manner that assures a connection between STEM and environmental education. A potential adjustment to the intervention to promote this is the inclusion of explicit guidance on reflecting upon these connections. This could be included in the final report, however Ring et al. (2017) utilised collaborative reflection and negotiation as an effective tool to promote conceptual change in integrated STEM. The addition of collaborative reflection may not only deepen PSTs understanding of integrated STEM but may help them reveal the underlying connections between STEM and EE.

Third, the assessment needs to be redesigned in such a way that it more explicitly targets these conceptions and connections that we have identified as insufficiently emphasised in this first iteration of the intervention (e.g. specific content understanding and connections between STEM and EE). The assessment criteria of the current intervention rubric is too focused on the design process, and does not equally represent the other aspects of the intervention which we wanted to highlight.

The action research approach used in this study has enabled us to reveal not only successful aspects of the E-STEM intervention, but it has also allowed us to reflect upon the ways that we can improve and add to the intervention so as to target areas that will result in future iterations of the intervention providing more valuable experiences for PSTs.

Acknowledgements This research was supported by a grant from the E. Wayne Gross Fund.

Appendix A: STEM Attitudes Survey Items with Corresponding Constructs

All items were scored on a 1–5 Likert scale (1 strongly disagree to 5 strongly agree). This survey is based on The BRAINS survey (Summers & Abd-El-Khalick, 2018).

Construct	Survey Item
Attitude	I enjoy STEM STEM is one of the most interesting topics to learn about I really enjoy learning about STEM I would like to do STEM activities at home I really like STEM I do not like STEM
Intention	I will study STEM during my time at university I will not pursue a STEM related career in the future I will become a STEM professional in the future I will continue to study STEM after I leave school I would enjoy working in a STEM related field I will take more STEM related classes in the future
Behaviour	STEM professionals are highly respected Most people should understand STEM because it affects their lives We live in a better world because of STEM Knowing STEM can help me make better choices about my health Knowledge of STEM helps me protect the environment STEM will help me understand the world around me STEM professionals usually like to work even if they have a day off People with STEM related careers have a normal family life I am encouraged to understand concepts when learning about STEM
Control	I am sure I could do well on a STEM test I usually give up if I do not understand a STEM concept STEM is easy for me I cannot understand STEM even if I try hard I can understand difficult STEM concepts I am confident that I can understand STEM
Normative	Members of my family work in STEM fields My family encourages my interest in STEM My family encourages me to have a STEM related career

Appendix B: Report Assessment Rubric

	Outstanding	Good	Adequate	Inadequate
Identification and analysis 10	Clearly identifies and explains the purpose of the task Uses information to situate the task in real-world contexts	Clearly identifies and explains the purpose of the task	Identifies and explains the purpose of the task but explanations lack clarity	Does not identify or explain the purpose of the task
Design justification 30	Applies prior knowledge and data effectively to justify design choices. Justification is scientifically accurate. Design adjustments are discussed and justified in light of data and scientific concepts. Data is presented appropriately. Accurately defines where and how STEM has been used in order to complete the task	Applies prior knowledge and data effectively to justify design choices. Justification is scientifically accurate. Data is presented appropriately. Design adjustments are discussed but not linked to data or scientific concepts. Accurately defines where and how STEM has been used in order to complete the task	Applies prior knowledge somewhat effectively to justify design choices. Data is utilized in order to determine design adjustments. Somewhat defines where and how STEM has been used in order to complete the task	Prior knowledge is not used effectively to justify design choices. Data is not used effectively to justify design choices. Inadequately defines how and where STEM has been used in order to complete the task

	Outstanding	Good	Adequate	Inadequate
Identify future modifications 30	Students are able to identify components and/or elements of the design that can be further modified in order to improve the design of the model. Students are able to provide accurate scientific reasoning using appropriate scientific concepts to support the proposed modification of the components/ elements of the model. Students use outside sources in order to support their proposed modifications. Sources are cited using proper APA formatting	Students are able to identify components and/ or elements of the design that can be further modified in order to improve the design of the model. Students support their proposal using limited scientific reasoning. Students use outside sources to support their proposed modifications and cite them using proper APA formatting.	Students are able to identify components and/ of elements of the design that can be further modified to improve the design of the model. Scientific reasoning for the modifications is limited.	Element/ component modification suggestion are not present and/or are not justified using scientific concepts or reasoning.
Conclusion 10	Conclusion coherently summarizes the current design and suggestions for further exploration of design/material modifications	Conclusion coherently summarizes the current design and suggestions for further exploration of design/material modifications	Conclusion coherently summarizes the current design and suggestions for further exploration of design/material modifications	Conclusion coherently summarizes the current design and suggestions for further exploration of design/material modifications
References 10	Correct APA formatting used for in-text citations and in a complete reference list.	Correct APA formatting used for in-text citations and in a complete reference list.	Correct APA formatting used for in-text citations and in a complete reference list.	Correct APA formatting used for in-text citations and in a complete reference list.
Formatting and grammar 10	Appropriate grammar, punctuation and spelling	Appropriate grammar, punctuation and spelling	Appropriate grammar, punctuation and spelling	Appropriate grammar, punctuation and spelling

References

Adams, A. E., Miller, B. G., Saul, M., & Pegg, J. (2014). Supporting elementary pre-service teachers to teach STEM through place-based teaching and learning experiences. *Electronic Journal of Science Education, 18*(5), n5.

Ajzen, I. (1985). From intentions to actions: A theory of planned behavior. In *Action control* (pp. 11–39). Berlin, Germany\Heidelberg, Germany: Springer.

Akerson, V. L., Burgess, A., Gerber, A., Guo, M., Khan, T. A., & Newman, S. (2018). Disentangling the meaning of STEM: Implications for science education and science teacher education. *Journal of Science Teacher Education, 29*(1), 1–8.

Anderson, V., Datta, R., Dyck, S., Kayira, J., & McVittie, J. (2016). Meanings and implications of culture in sustainability education research. *The Journal of Environmental Education, 47*(1), 1–18.

Awad, N., & Barak, M. (2018). Pre-service science teachers learn a science, technology, engineering and mathematics (STEM)-oriented program: The case of sound, waves and communication systems. *Eurasia Journal of Mathematics, Science and Technology Education, 14*(4), 1431–1451.

Barnett, S. M., & Ceci, S. J. (2002). When and where do we apply what we learn?: A taxonomy for far transfer. *Psychological Bulletin, 128*(4), 612.

Bell, D. R. (2004). Creating green citizens? Political liberalism and environmental education. *Journal of Philosophy of Education, 38*(1), 37–54.

Berry, A., McLaughlin, T., & Cooper, G. (2018). Building STEM self-perception and capacity in pre-service science teachers through a school-university Mentor program. In *STEM education: An emerging field of inquiry* (pp. 190–207). Leiden, The Netherlands: Brill Sense.

Birney, L., & Cronin, J. (2019). Environmental habitat restoration and inquiry-based learning with New York City public schools—an urban model in STEM education. *Journal of Environmental Studies and Sciences, 9*(3), 322–326.

Blackley, S., & Howell, J. (2015). A STEM narrative: 15 years in the making. *Australian Journal of Teacher Education, 40*(7), 8.

Bonnett, M. (2002). Education for sustainability as a frame of mind. *Environmental Education Research, 8*(1), 9–20.

Breiner, J. M., Harkness, S. S., Johnson, C. C., & Koehler, C. M. (2012). What is STEM? A discussion about conceptions of STEM in education and partnerships. *School Science and Mathematics, 112*(1), 3–11.

Bruner, J. S. (1966). *Toward a theory of instruction* (Vol. 59). Cambridge, MA: Harvard University Press.

Bybee, R. W. (2013). *The case for STEM education: Challenges and opportunities*. Arlington, VA: NSTA Press.

Cranton, P. (2006). *Understanding and promoting transformative learning: A guide for educators of adults* (2nd ed.). San Fransisco: Jossey-Bass.

Dewey, J. (1986, September). Experience and education. In *The Educational Forum* (Vol. 50, No. 3, pp. 241–252). Taylor & Francis Group.

Eilks, I. (2018). Action research in science education: A twenty-year perspective. *ARiSE, 1*(1), 3–14.

Eilks, I., & Ralle, B. (2002). Participatory Action Research within chemical education. *Research in chemical education-what does this mean* (pp. 87–98).

English, L. D. (2016). STEM education K-12: Perspectives on integration. *International Journal of STEM Education, 3*(1), 3.

Feldman, A. (1996). Enhancing the practice of physics teachers: Mechanisms for the generation and sharing of knowledge and understanding in collaborative action research. *Journal of Research in Science Teaching, 33*(5), 513–540.

Festinger, L. (1957). *A theory of cognitive dissonance* (Vol. 2). Stanford university press.

Fraser, J., Gupta, R., Flinner, K., Rank, S., & Ardalan, N. (2013). *Engaging Young People in 21st Century Community Challenges: Linking Environmental Education with Science, Technology, Engineering and Mathematics*. New York: New Knowledge Organization.

Glaser, B. G. (1978). *Strauss (1967): The discovery of grounded theory: Strategies for qualitative research* (Vol. 81, p. 86). London: Wiedenfeld and Nicholson.

Grundy, S. (1982). Three modes of action research. *Curriculum Perspectives, 2*(3), 23–34.

Guzey, S. S., Moore, T. J., & Harwell, M. (2016). Building up STEM: An analysis of teacher-developed engineering design-based STEM integration curricular materials. *Journal of Pre-College Engineering Education Research (J-PEER), 6*(1), 2.

Holdsworth, S., & Thomas, I. (2016). A sustainability education academic development framework (SEAD). *Environmental Education Research, 22*(8), 1073–1097.

Honey, M., Pearson, G., & Schweingruber, H. A. (Eds.). (2014). *STEM integration in K-12 education: Status, prospects, and an agenda for research* (p. 180). Washington, DC: National Academies Press.

Howlett, C., Ferreira, J. A., & Blomfield, J. (2016). Teaching sustainable development in higher education: Building critical, reflective thinkers through an interdisciplinary approach. *International Journal of Sustainability in Higher Education, 17*(3), 305–321.

Jones, P., Selby, D., & Sterling, S. (2010). More than the sum of their parts? Interdisciplinarity and sustainability. In *Sustainability education: Perspectives and practice across higher education* (pp. 17–38). Washington, DC: Earthscan.

Kelley, T. R., & Knowles, J. G. (2016). A conceptual framework for integrated STEM education. *International Journal of STEM Education, 3*(1), 11.

Kennedy, T. J., & Odell, M. R. L. (2014). Engaging students in STEM education. *Science Education International, 25*(3), 246–258.

Kopnina, H. (2012). Education for sustainable development (ESD): The turn away from 'environment' in environmental education? *Environmental Education Research, 18*(5), 699–717.

Lave, J., & Wenger, E. (1991). *Situated learning: Legitimate peripheral participation*. Cambridge, UK: Cambridge University Press.

Merritt, E., Hale, A., & Archambault, L. (2019). Changes in pre-service teachers' values, sense of agency, motivation and consumption practices: A case study of an education for sustainability course. *Sustainability, 11*(1), 155.

NAAEE. (2020, June 9). *E-STEM*. NAAEE. https://naaee.org/our-work/programs/e-stem

Nadelson, L. S., Callahan, J., Pyke, P., Hay, A., Dance, M., & Pfiester, J. (2013). Teacher STEM perception and preparation: Inquiry-based STEM professional development for elementary teachers. *The Journal of Educational Research, 106*(2), 157–168.

Ostler, E. (2012). 21st century STEM education: A tactical model for long-range success. *International Journal of Applied Science and Technology, 2*(1).

Pitt, J. (2009). Blurring the boundaries–STEM education and education for sustainable development. *Design and Technology Education: An International Journal, 14*(1).

Ring, E. A., Dare, E. A., Crotty, E. A., & Roehrig, G. H. (2017). The evolution of teacher conceptions of STEM education throughout an intensive professional development experience. *Journal of Science Teacher Education, 28*(5), 444–467.

Rinke, C. R., Gladstone-Brown, W., Kinlaw, C. R., & Cappiello, J. (2016). Characterizing STEM teacher education: Affordances and constraints of explicit STEM preparation for elementary teachers. *School Science and Mathematics, 116*(6), 300–309.

Rodgers, C. (2002). Defining reflection: Another look at John Dewey and reflective thinking. *Teachers College Record, 104*(4), 842–866.

Shernoff, D. J., Sinha, S., Bressler, D. M., & Ginsburg, L. (2017). Assessing teacher education and professional development needs for the implementation of integrated approaches to STEM education. *International Journal of STEM Education, 4*(1), 13.

Slavit, D., Nelson, T. H., & Lesseig, K. (2016). The teachers' role in developing, opening, and nurturing an inclusive STEM-focused school. *International Journal of STEM Education, 3*(1), 7.

Stevenson, R. B., Brody, M., Dillon, J., & Wals, A. E. (2013). *International handbook of research on environmental education*. New York: Routledge.
Stohlmann, M., Moore, T. J., & Roehrig, G. H. (2012). Considerations for teaching integrated STEM education. *Journal of Pre-College Engineering Education Research (J-PEER), 2*(1), 4.
Summers, R., & Abd-El-Khalick, F. (2018). Development and validation of an instrument to assess student attitudes toward science across grades 5 through 10. *Journal of Research in Science Teaching, 55*(2), 172–205.
Tilbury, D. (2011). Higher education for sustainability: A global overview of commitment and progress. *Higher Education in the World, 4*, 18–28.
Vare, P., & Scott, W. (2007). Learning for a change: Exploring the relationship between education and sustainable development. *Journal of Education for Sustainable Development, 1*(2), 191–198.
Vygotsky, L. (1978). Interaction between learning and development. *Readings on the Development of Children, 23*(3), 34–41.
Wals, A. E. J., & Jickling, B. (2002). "Sustainability" in higher education: From doublethink and newspeak to critical thinking and meaningful learning. *International Journal of Sustainability in Higher Education, 3*(3), 221–232.

Angela Burgess is a Doctoral Student of Science Education at Indiana University and an informal environmental educator. Her research focuses on environmental literacy, environmental education practices and participant outcomes.

Gayle A. Buck is an Associate Dean for Research, Development & Innovation, as well as a Professor of Science Education. Previously a middle level science teacher in both urban and rural schools, Professor Buck now teaches courses in science, STEM education and teacher education. Her research explores (1) student populations traditionally underserved in science education, (2) neglected epistemological assumptions in teaching and learning, and (3) pragmatic and participatory approaches to educational research.

Chapter 6
Navigating Theory and Practice: Digital Video Games (DVGs) in STEM Education

Isha DeCoito and Lisa K. Briona

6.1 Introduction

Globally, education policy reform has called for preK-16 schools to improve students' twenty-first century learning skills. In particular, the emphasis is on deeper integration of numeracy, scientific and technological literacy within the curriculum, and their understanding of science and technology content, socio-scientific issues, the nature of science, and scientific and technological problem-solving (DeCoito, 2017; Millar, 2006). Additionally, in this rapidly changing and technologically evolving world there is a need for skilled labour and professionals in STEM fields. An emphasis on STEM stresses a multidisciplinary approach to better prepare all students in STEM subjects and increase the number of postsecondary graduates who are prepared for STEM occupations (Conference Board of Canada, 2013; NRC, 2011). Students must be inspired, engaged, and have deep understandings of STEM content and their applications if they are to consider future studies and/or jobs in STEM fields (DeCoito, 2016). This warrants concern as the number of individuals graduating from and/or pursuing careers in STEM fields is seriously lagging (Mishagina, 2012). When available, today's students can leverage suitable STEM education and training to address the STEM labour and skills gap and secure their futures (Abdalla et al., 2018).

To support successful adaptation to the ever-increasing integration of technology into our lives, it is vital that education emphasize technological literacy, especially for students intending to pursue STEM focused careers, such as engineering, architecture, medicine or information technology. One avenue for enhancing scientific, engineering, technological and numeracy literacies, while at the same time

I. DeCoito (✉) · L. K. Briona
Western University, London, Canada
e-mail: idecoito@uwo.ca

V. L. Akerson, G. A. Buck (eds.), *Critical Questions in STEM Education*, Contemporary Trends and Issues in Science Education 51,
https://doi.org/10.1007/978-3-030-57646-2_6

supporting the development of twenty-first century skills, is through the implementation of digital video games (DVGs) within an integrated STEM education context.

In this chapter, the authors discuss the use of DVGs as a learning environment; one that engages learners through technology, provides opportunity for creative output, and promotes the intrinsic motivation for learning necessary to advance the development of twenty-first century skills and interaction with STEM content. The authors highlight the development of DVGs in a STEM methods course in teacher education along with teacher candidates' reflections on DVG development, and subsequent incorporation of DVGs in a practicum setting by one of the authors.

6.1.1 STEM Education

A major goal of the STEM agenda is to improve the proficiency of all students in STEM, regardless of whether or not they choose to pursue STEM careers or post-secondary studies, while at the same time fostering the twenty-first century skills identified as being crucial for success (Orpwood, Schmidt, & Jun, 2012). These skills include critical thinking, problem solving, creativity, collaboration, self-directed learning, as well as scientific, environmental, and technological literacies (Howard-Brown & Martinez, 2012). STEM education is a rethinking of traditional approaches to teaching STEM subjects, whereby these four strands are integrated into one "meta-discipline" (Fioriello, 2010). While many definitions of STEM exist, we are focusing on an approach that stresses meaningful interdependence among all disciplines of STEM. In other words, including all individual disciplines of STEM (science, technology, engineering, and mathematics) in a way that is meaningful and showcases the interdependence of the fields. Despite the various definitions of STEM, educators struggle with integrating these disciplines for various reasons, including content knowledge, pedagogical knowledge, and pedagogical content knowledge (DeCoito & Myszkal, 2018; Koehler & Mishra, 2009), to name a few.

Successful STEM education goes hand-in-hand with effective instructional practices. Few teachers are prepared to operationalize STEM education, and according to Wilson (2011), better preparation of students lies in effective instruction that actively engages them in science, mathematics, and technology engineering practices throughout their schooling, and broadens their awareness of STEM careers. Moreover, research has demonstrated that integrated STEM education is student-centered, can increase retention and promote problem solving and higher-level thinking skills (Stohlmann, Moore, & Roehrig, 2012). In this way, students successively deepen their understanding, both of core ideas in the STEM fields and of concepts that are shared across areas of STEM (DeCoito & Myszkal, 2018).

6.1.2 Teacher Education

The task of preparing teachers to become technologically competent is challenging and requires ample effort, time, and opportunities. Some teachers do not feel prepared to integrate technology into their practice; not due to a lack of the technology itself, but rather due to a lack of knowledge of how to integrate the technology into their teaching. Ruggiero and Mong (2015) described factors that potentially influence teachers' use of technology as both external (resources, training, and support) and internal (personal investment in technology, attitude toward technology, and peer support) barriers. Often, a lack of comfort by teachers in teaching the required content is reflected in poor preparation of students (Schmidt, 2011). Weak initial teacher preparation heightens the importance of continuing professional development, but the available research suggests that professional development in STEM, when available, is often short, fragmented, ineffective, and not designed to address the specific need of individual teachers (Wilson, 2011). One area that warrants much attention is teacher education programs, as they can potentially impact STEM education in a positive manner. For example, in a study focused on enhancing science teacher candidates' (TC) digital competencies, DeCoito (2017) found that that the explicit integration of digital literacies created and engaged learning communities, while improving technological and scientific literacies in a purposeful manner.

In general, teacher education programs are lacking in terms of effectively modeling technology integration into practice. According to Gray, Thomas, and Lewis (2010), reporting on the National Center for Education Statistics 2009 survey of US teachers, 75% felt that undergraduate teacher education programs and 67% felt that graduate teacher education programs failed to prepare them to use educational technology effectively for instruction. Alarmingly, 78% indicated that they learned to use technology in their classrooms, independent of teacher education programs. This is problematic, given the ever-expanding role of technology in society, and the changing face of curriculum. It speaks directly to necessary changes in teacher education programs such that the roles of teachers are re-envisioned to support learners in being proactive about, and active participants in their learning (Bolstad et al., 2012) and their future participation in STEM professions.

6.1.3 Digital Literacies and DVGs

A promising new frontier lies in using games-based learning to better prepare students for careers in STEM-related fields. For example, DeCoito (2014) explored digital scientific timelines to teach about the nature of science, and more recently, the potential of digital games to enhance scientific and technological literacy while teaching STEM concepts (DeCoito & Richardson, 2016). Whether STEM skills are taught through the content of digital games or by building a DVG from scratch, one thing is clear: DVGs are a powerful force in young people's lives. Based on the manner in which DVGs are introduced into education, they have the potential to effectively integrate STEM concepts and principles. Games have the remarkable ability to engage and motivate players, often just through the joy of playing (Dicheva, Dichev, Agre, & Angelova, 2015). In an educational setting, several researchers have reported that students are more motivated to learn gamified concepts, and that they consider them both easier to learn, and easier to recall in test settings (Barata, Gama, Jorge, & Gonçalves, 2013; Byl & Hooper, 2013).

Despite today's students being fully engaged with DVGs as vehicles for edutainment, very few educators use DVGs in any substantive way for teaching and learning (Annetta, 2008; DeCoito & Richardson, 2016). Video games can enable STEM education from elementary school to post-secondary as they teach both hard and soft skills such as analytical thinking, multitasking, strategizing, problem-solving, and team building. Video games represent the kind of interactive and self-paced learning that people see as a future guided by technology, with the caveats that games cannot replace traditional teaching methods, and games should not represent the sole teaching/learning component. In spite of games providing opportunities for self-learning, students also need guidance and mentorship from their teacher-as-facilitator. The role of the teacher here is not diminished but becomes more challenging and interesting in terms of scaffolding (Vygotsky, 1978) students to learn with this technology.

In addition to addressing a variety of STEM skills, DVGs can potentially address seven of the eight practices of science and engineering, identified in *A Framework for K-12 Science Education* (NRC, 2011) as essential for all students to learn. For example, DVG development can engage students in: i) asking questions (for science) and defining problems (for engineering); ii) developing and using models; iii) using mathematics, information and computer technology, and computational thinking; iv) constructing explanations (for science) and designing solutions (for engineering); v) engineering design (all components); vi) linking engineering, technology, science and society; and vii) obtaining, evaluating, and communicating information.

6.1.4 The Engineering Design Process (EDP) and DVG Design

Engineering is frequently considered the most difficult aspect of STEM to implement in a K-12 classroom setting, and many have argued that it is not possible to teach anything approximating "real engineering" without advanced math and science knowledge (Bush et al., 2006; Roehrig, Moore, Wang, & Park, 2012). However, part of this perceived difficulty lies in a misunderstanding in what engineers actually do, and a general lack of access to engineers: how often do we get to meet the civil engineer that designed the bridge we drive across on our way to work or school every morning? By comparison, lay people have a good general grasp of medicine because of its pervasiveness in our daily lives: many television shows feature actors portraying medical professionals; we have regular interactions with doctors, nurses; etc.

The EDP process differs from the scientific process in a very important way. While both involve research and testing, the scientific process involves creating a hypothesis or theory about the world. The EDP is a problem-solving activity, requiring a step-by-step iterative methodology whereby a design problem and its solution(s) co-evolve (Dym & Little, 2000). The process is considered an authentic learning experience in which students learn by doing.

Storyboarding is a key aspect of the EDP, since storyboards lay the foundation for DVGs, outlining characters, scenes, challenges, and gaming interface. In a professional setting, storyboards also help prevent "scope creep," where continuous changes are implemented, or additional expectations are added, causing a potential budget to overrun and missed milestones/ deadlines (Thakurta, 2013). An example of scope creep from the construction field might be an enthusiastic first-time home builder who initially approves a set of plans and budget, then after construction has commenced requires that the kitchen cabinets be moved to the opposite wall, the unfinished basement be finished and contain a family room, wet bar and fireplace, and that the back patio be converted into an enclosed three-season sunroom.

In DVG design, storyboards are frequently accompanied by design documents, which track aspects of game play including leveling/scoring rules, plot progression, character backstory and key interactions, unlockable features, and "Easter eggs" – intentional hidden messages or secret rooms placed within a game to encourage and reward game play. Creating smaller games where you are the designer, animator, and programmer offers the luxury of treating them as living documents that can be responsive to constraints, limitations, and new ideas.

The American Society for Engineering Education (2011) notes that, "engineers solve problems using science and math." Truesdell (2014) has observed that the ability to use "clear and concise problem formulation," that is, to define a project sufficiently to prevent scope creep, is critical to an engineer's ability to identify a best solution. Together, these observations strongly support the claim that storyboarding, and design documentation, *is* engineering.

The following sections provide a framework for developing DVGs as a strategy for integrating the four STEM disciplines, while at the same time providing opportunities for TCs to enhance STEM literacy, problem solving skills, STEM content knowledge, self-efficacy, coding and computational thinking skills, and career awareness in STEM.

6.2 DVGs in Teacher Education

6.2.1 Context

The development of DVGs being reported is an assignment situated in a curriculum and pedagogy course focusing on STEM education within the broader curricular spectrum. In order to connect video games to the world of STEM learning, TCs were required, over four 2-hour seminars, to develop a DVG to teach STEM concepts in biology, chemistry or physics. TCs were provided opportunities to develop a variety of STEM skills while creating their DVGs, including programming, math, creative thinking, logic, and using the EDP. One of the main goals of the assignment, in addition to incorporating STEM content knowledge, was to introduce TCs to the fundamentals of coding and computational thinking, which are critical skills for programmers and developers. As with any language, it is highly effective to learn by doing; thus, the assignment is an example of a real-world experience that engages students in STEM learning. In addition to the aforementioned goal and given the inequitable access to technological resources in schools, this assignment played a vital role in i) providing opportunities for teachers to engage in the process of developing educative materials linked to curriculum, ii) preparing them for using and modeling technology effectively in their future practices by enhancing and expanding their skill sets, and iii) engaging in best practices associated with integrated STEM education.

TCs were familiarized with DVGs through discussions of selected literature on games-based learning. Thereafter, TCs participated in two workshops led by a guest speaker. The first workshop introduced DVGs through a practitioner's lens, in which the guest speaker discussed a computer programming/physics-related career project conducted by grade 12 students to teach grade 10 students about different careers in physics reported elsewhere. TCs then engaged in exploring curricular areas that would benefit from the inclusion of DVGs and were reminded that games should be used in conjunction with good teaching. A number of programming languages, platforms, and engines, including Eclipse (https://eclipse.org/), Hour of Code (https://code.org/learn), and Beyond the Hour of Code (https://code.org/learn/beyond) were discussed. In preparation for the second workshop, TCs were required to explore a programming language and select a concept from the curriculum well-suited for DVG integration. During the second workshop, key aspects of computer game design and different types of games were explored: action, adventure, role-playing, simulation, strategy, skill-based, and other types, along with benefits and challenges (Richardson & DeCoito, 2017).

The instructor introduced the EDP and asserted this as a preferred framework for developing the DVGs. This process, which is a key aspect of the *Next Generation Science Standards*, provided TCs with opportunities to develop a sophisticated problem-solving process, that is transferrable to other disciplines.

6.2.2 DVG Criteria

Key criteria considered imperative to the DVG research and development were: i) a STEM education focus, and ii) DVG design. In the former, curricular connections were to be explicit and include three of the four STEM subjects. This choice was provided given the diversity of teacher candidates' subject matter expertise and interest. In DVG design, career connections were embedded and a pluriversal approach was employed. DVG criteria included mandatory elements: rewards, levelling, and an avatar. The presence of an avatar promotes intrinsic motivation to achieve the game objectives, facilitates game immersion, and enhances the game play experience (Birk, Atkins, Bowey, & Mandryk, 2016). Additionally, the ability to name and customize an avatar supports personalized learning and game investment (Konstantinidis, Tsiatsos, & Pomportsis, 2009).

To monitor progress, TCs were instructed to storyboard their game and submit reflections based on the process of researching and developing their DVG, up to and including the storyboarding phase. Feedback was provided by the instructor before TCs could move to the next stage of the DVG development. TCs were provided in-class opportunities to collaborate and receive online mentorship. Finally, the completed DVGs were submitted, along with TCs' collective reflections on developing DVGs. TCs were encouraged to implement their DVG and/or DVG development in their practicum classrooms. TCs' reflections will be discussed later in the chapter, as well as reflections on the modeling and implementation of a similar DVG activity with high school students.

The following section highlights *Okazaki's Revenge* (http://OkazakisRevenge.GamifiedLearningRD.com), a DVG developed by one of the authors who was a TC during the assignment. It is noteworthy to mention that this particular DVG incorporates all four STEM subjects. This example guides the reader through the game scenario and mechanics, storyboarding and design, and includes a game walkthrough.

6.2.3 DVG: Okazaki's Revenge

Many students in senior-level biology classes struggle with understanding prokaryotic DNA replication (Knippels, Arend Jan Waarlo, & Boersma, 2005; Reddy, Bhaskara, & Mint, 2017). As the Ontario curriculum requires students to use appropriate terminology related to molecular genetics, and to be able to explain the current model of DNA replication (Ontario Ministry of Education, 2008), one of the authors sought to make these concepts accessible using a gamified approach.

Game Scenario and Mechanics Loosely inspired by the 1987 movie *Innerspace*, *Okazaki's Revenge* is set in the year 3067: human cancers no longer exist because we rely on nanobots to both ensure our DNA replication accuracy and to destroy any emergent cancer cells before they gain a foothold in our bodies. Players assume the role of a technician trainee responsible for the actions of one of these nanobots.

Within the game, nanobots serve two functions: i) to ensure that enzymes associated with DNA replication are placed at the correct location on DNA strands and in the correct temporal sequence, and ii) to destroy the cancer cells before they overrun the host if a cell fails to complete DNA replication correctly and thus turns cancerous. All aspects of the game – 'History of DNA 101,' 'DNA Replication,' and game play – highlight the science and mathematics components of STEM.

While the game interface allows players to go directly to the scored challenge, there is an accompanying two-part teaching module that introduces the basics of DNA and prokaryotic DNA replication, which is accessible at any time of game play. There is also a guided practice arena where players are provided visual cues and enzyme names to learn the order of DNA replication. The practice arena is a standard game mechanism employing aspects of Vygotsky's zone of proximal development (1978) in that it provides the player a safe environment to learn new key combinations or complex movement routines without penalty. By giving players the opportunity to gain proficiency without incurring punishment, they are challenged but not frustrated to the point of quitting.

Game Walkthrough After the initial welcoming splash screen, players are invited to select an avatar (Fig. 6.1). While the teacher modules (DNA Basics & DNA Replication) can be accessed without choosing an avatar, the practice arena and game play field are disabled until one is established. Once the avatar is named, players are situated with the game's plot.

The 'DNA Basics' are seven screens that introduce: i) the discovery of DNA by Friedrich Miesher (Fig. 6.2, *left*); ii) the elucidation by Phoebus Levene that DNA consists of a deoxyribose sugar, a phosphate group, and a nitrogenous base; iii) the structures and classification of guanine, adenine, cytosine, and thymine; iv) hydrogen bond formation between purines and pyrimidines as characterized by Erwin Chargaff; v) numbering the carbons in a molecule of deoxyribose; vi) how a nucleotide is formed from the principle constituents (Fig. 6.2, *right*); and vii) the contributions that Rosalind Franklin, James Watson, and Francis Crick made to our understanding of DNA organization, and that DNA consists of two antiparallel strands normally written in a 5′ → 3′ format. These slides review the science of genetics in an interactive manner (click-based navigation) supporting a familiar passive learning style (reading-only, with no responsive feedback). The writing is

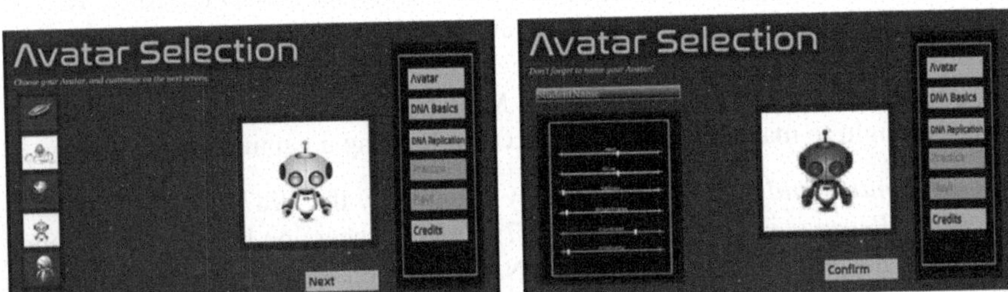

Fig. 6.1 *Left:* Choosing an avatar in *Okazaki's Revenge*. *Right*: Customizing the avatar

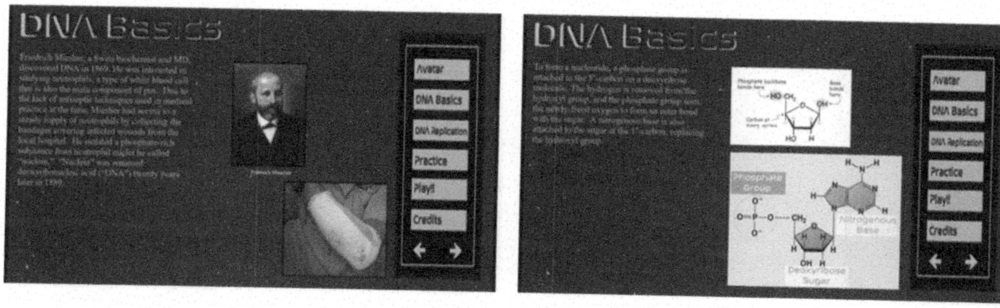

Fig. 6.2 *Left*: DNA Basics slide 1 – the discovery of DNA. *Right*: DNA Basics slide 6 – the formation of a nucleotide from the principle constituents

simple and accessible and provides definitions and clarifications that address common misconceptions associated with genetics (Mills Shaw, Van Horne, Zhang, & Boughman, 2008). Portraits of all scientists mentioned humanizes the science, and clearly labeled figures promotes understanding.

During *Okazaki's Revenge* gameplay the discipline of mathematics is relevant to the structure of DNA. Geometry describes the shape of an object at a particular instance in time. The most common representation of DNA in high school textbooks is of the B-form: two strands of DNA coiling around each other, with the minor groove of one seamlessly fitting into the major groove of the other, and each helix mathematically defined using simple trigonometry. Topology describes how an object deforms, such as how double-stranded DNA unzips during replication. Using the observations that the distance between base pairs is ~3.4 Å, the diameter of B-DNA is ~20 Å, and that each turn of the helix is about 10 base pairs, biologists can readily calculate the length of a helix (~34 Å), or the number of base pairs in a given length of DNA. While these are simple calculations, they are nontrivial, and are relevant to more advanced topics such as molecular biology and synthetic biology. In addition to mathematics being highlighted in DNA structure, the logical progression of learning and the rules of gameplay in *Okazaki's Revenge* necessitates an understanding of mathematical concepts to develop and implement the DVG successfully (discussed later).

'DNA Replication' is a series of 10 slides that navigate through prokaryotic DNA replication, accompanied by looping animations. Facts are highlighted about DNA replication including: i) it is semi-conservative; ii) it begins at the origin of replication (Fig. 6.3, *left*); iii) uses DNA helicase to break the H-bonds between complementary base pairs; iv) it employs single-stranded binding proteins to prevent premature re-annealing; v) it recruits DNA gyrase to eliminate supercoiling; vi) it uses primase to attach an RNA primer prior to replication; vii) it engages Pol III to replicate the template strand of DNA (Fig. 6.3, *right*); viii) it uses Pol I to remove and replace the RNA primers; ix) it has ligase join the Okazaki fragments; and x) it uses Poll II as a proof reader.

The practice arena (Fig. 6.4, *left*) has an identical layout to the gameplay screen (Fig. 6.4, *right*). In the practice arena, the game layout is explained using the two

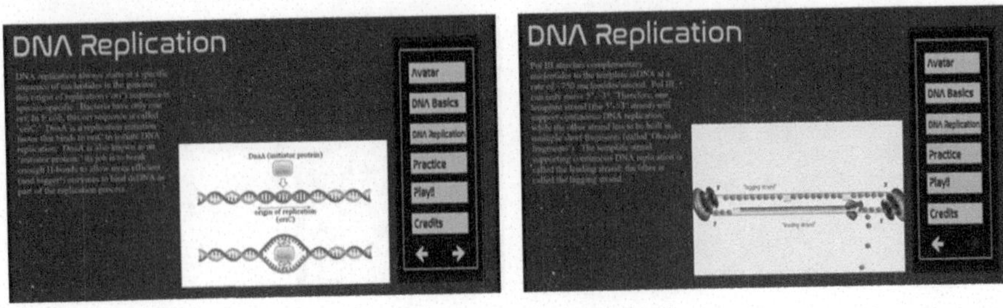

Fig. 6.3 *Left*: DNA Replication slide 2 – the origin of replication. *Right*: DNA Replication slide 7 – engaging Pol III to replicate the template strand of DNA

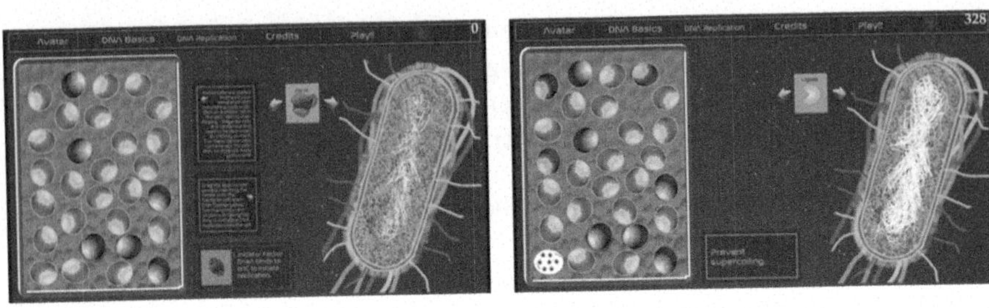

Fig. 6.4 *Left*: Practice arena. *Right*: Game play in which players choose a dark grey cell from the petri dish, then select and apply the correct protein/enzyme based on the prompt in the box

instructional boxes in the centre of the screen, while hints are provided in the lower box. Players are instructed and prompted to become proficient in facilitating DNA replication. When the bacterial cell on the right requires nanobot assistance during the DNA replication process, the DNA (white) flashes within the cell (compare the bacterium cross-section in Fig. 6.4, *left*, with Fig. 6.4, *right*). The player can cycle through the various proteins/enzymes associated with DNA replication using the left and right arrows. In both practice and live gameplay, the proteins/enzymes are named. However, only in the practice arena are players prompted with hints (bottom middle of screen). Additionally, during practice players have an unlimited amount of time to find the correct protein/enzyme to apply to the cell on the right; during gameplay they have 30 seconds before replication fails and the cell "dies." Both screenshots shown in Fig. 6.4 include a petri dish of cells (shown on the left side of the game interface) that the technician is responsible for maintaining. Light grey cells are in interphase or undergoing DNA replication without incident, and require no intervention. Dark grey cells are stalled in the DNA replication process and require assistance (in the petri dish shown in Fig. 6.4, *left*, there are six cells requiring intervention). Players choose which dark grey cell to "fix," as each cell is randomly assigned the replication step at which it is stalled. During practice and level 1 of gameplay, dark grey cells can accumulate without penalty. At level 2 of gameplay, however, if they remain dark grey too long (90 seconds), they become speckled, signifying that they have become cancerous (Fig. 6.4, *right*). Players "destroy" cancer cells by clicking on them. Gameplay is over when the petri dish becomes overrun with cancerous cells.

Game Storyboard and Design Technology was utilized in the development and implementation of *Okazaki's Revenge*. Construct 2 (www.scirra.com) was employed as an HTML5-based visual game editor to create *Okazaki's Revenge* for both its operating system independence (games created with Construct 2 run on any platform, and in any HTML5-compliant browser), and to evaluate its efficacy at novice/non-programmer support. Computation thinking (CT), defined as "the thought process involved in formulating a problem and expression its solution(s) in such a way that a computer – human or machine – can effectively carry out" (Wing, 2014), is inherent in the design of *Okazaki's Revenge*. CT involves problem solving (via logical thinking, such as deciding whether or not a particular problem can be solved computationally), building algorithms (describing step-by-step procedures to solve a particular problem), and execution (programming and debugging a solution). Only this last core skill can be addressed via programming.

The first core skill of CT – problem solving – involves defining the scope and parameters of a problem, and frequently includes problem decomposition: breaking down a complex problem into smaller problems that are simpler, more discrete and usually easier to solve. It also involves identifying the order in which these smaller problems need to be solved in order to solve the parent problem. Algorithm building describes a solution in terms of actions, loops, or decisions. Actions are a set of one or more instructions that are completed sequentially. For example, in the DVG *Okazaki's Revenge*, the welcome screen is displayed first, followed by the "Choose Avatar screen." These two sequential instructions represent an Action in the *Okazaki's Revenge* algorithm prototype. Loops are a set of instructions that are repeated as a whole. In the *Okazaki's Revenge* petri dish, each cell's movement path is defined as a set of repeating steps: move x pixels up and y pixels left, wait a little while, then move z pixels down and γ pixels right, wait a little while, and repeat. Decisions require a choice (e.g., players of *Okazaki's Revenge* choose an avatar from a selection of five). Another example of a decision is when players drag a protein/enzyme into the cell. If the correct enzyme is placed, then the score increases and the cell disappears; however, if an incorrect enzyme is placed, then the player loses 5 points and has to try again.

The initial storyboard and design document (Fig. 6.5) for *Okazaki's Revenge* uses a screenshot from the *How Nucleotides Are Added in DNA Replication* animation (Raven, 2007) to illustrate key game play content. The two images show the anticipated split-screen game play: the upper portion of the screen mimics a petri dish of healthy cells that are in interphase (mid-grey), undergoing mitosis (dark outline with light grey center), or cancerous cells needing destruction (dark grey). The lower portion of the screen shows a double-stranded DNA in the process of replication, and players would need to place the appropriate proteins/enzymes to help complete the replication process. Additionally, the design document identifies the game's purpose, curriculum correlation, scenario and scenes. It also describes which aspects of gameplay can be customized by the player, the levels, and the type(s) of gameplay employed. In the case of *Okazaki's Revenge*, adventure, first-person-shooter, and educational game motifs were utilized.

DVG Storyboard: DNA Replication

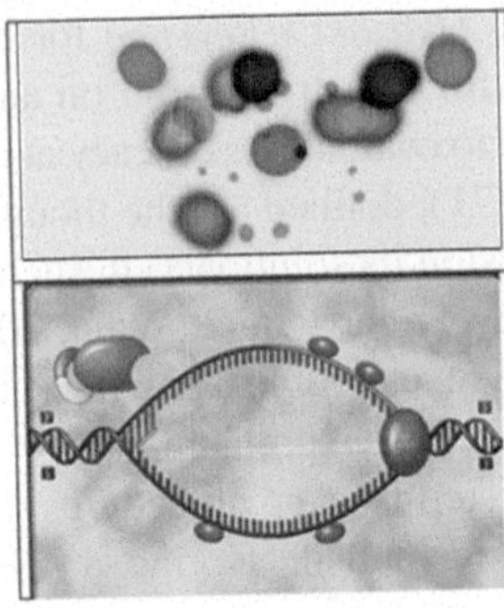

Figure 1: Main Game Screen

Goal: Players will be able to identify key enzymes and landmarks involved in prokaryotic DNA replication.

Strand: SBI4U, Molecular Genetics, Learning Objectives D2.1 and D3.1

Software: Construct2 (scirra.com)

Scenario: The year is 3067: cancer has been eradicated from our species due to utilization of nanobots to assist with enzymatic placement for DNA replication. Customize your 'bot, then get to work! But be warned: DNA replication must be completed accurately and timely, otherwise cancer cells arise. Destroy the cancer cells, lest they overrun your host.

Customization: Nanobot Avatar (colour, decorations)

User controls (left hand vs right hand setup)

Scenes: Welcome Screen
Setup (Avatar customization and control interface)
Teaching Module
Main Game
Game Over

Teaching Module: Walks players through steps of DNA replication and introduces them to the various keys combinations to effect an action (e.g. "w" key places DNA Polymerase I on DNA.

Game interface: Employs a vertical split-screen approach (Figure 1). The top part is a petri dish of cells going through replication: green cells are normal cells; blue cells are cells going through DNA replication and require player intervention to continue; magenta cells are cells that have failed to correctly complete DNA replication and are now cancerous. Magenta cells can be destroyed by clicking on them; blue cells can be clicked on to zoom in on their DNA in the lower part of the screen. If blue cell DNA replication is completed successfully, they turn green again; in the event of incorrect DNA replication they will turn magenta (cancerous).

Level 1: No cancer cells. Blue cells stay blue until DNA replication is successfully completed. Too many blue cells in the top screen (10?) = failed to complete the level. Successful completion of DNA replication for 10 cells lets the player move on to Level 2.

Level 2+: Cancer cells can now appear. Blue cells can appear faster also. If 50% of the cells in the petri dish are cancerous, then player failed to complete level (game over).

STEM: Science (Biology: DNA replication)
Technology (digital media; player is a nanobot)
Mathematics (modeling movement of interacting cell populations)

Career Options: Molecular Biology (biological processes)
Biophysics (artificial enzymes)
Analytical Chemistry (efficacy of enzymatic processes)
BioMathematics (modeling of cell populations)

Type of Game: action – strategy – simulation

Pluriversality: Western Modern Science: the various proteins and nucleotides involved in DNA replication
Ontological investment: do the cells in my petri dish live or die?

Fig. 6.5 Initial storyboard and design document for *Okazaki's Revenge*

Logic Flowcharting Game Mechanics Not only is logic a part of mathematics, it is the *language* of mathematics (Lawson, 2019; Roberts, 2010). Logic provides us with the ability to reason, to understand cause and effect, and to make rational decisions. In programming, logic flowcharts are a valuable part of developing the programming logic and game mechanics. They help programmers visualize gameplay and ensure all use-case scenarios are considered, beyond the obvious gameplay intention (e.g., what if the player doesn't ever choose a protein? What if there are no more light grey cells in the petri dish?). Figure 6.6 shows a simplified logic flowchart for the gameplay portion of *Okazaki's Revenge*. Diamond shapes represent points of decision, while rectangles represent actions. Logic flowcharts are typically written in the common tongue to maximize understanding when game development requires collaboration.

In most scored DVGs, gameplay becomes more challenging as the player gains experience and acumen. In *Okazaki's Revenge*, three levels are supported: Level One is for players with a score less than 500, Level Two is for players with a score between 501 and 1000, and Level Three is for players with a score greater than 1000. One way in which *Okazaki's Revenge* makes gameplay more challenging is via the reduction of time available to find the correct protein as the player progresses through the levels. At Level One, players have 30 seconds to find the correct protein when the DNA is flashing; at Level Two this decreases to 15 seconds, and at Level Three players have a mere eight seconds.

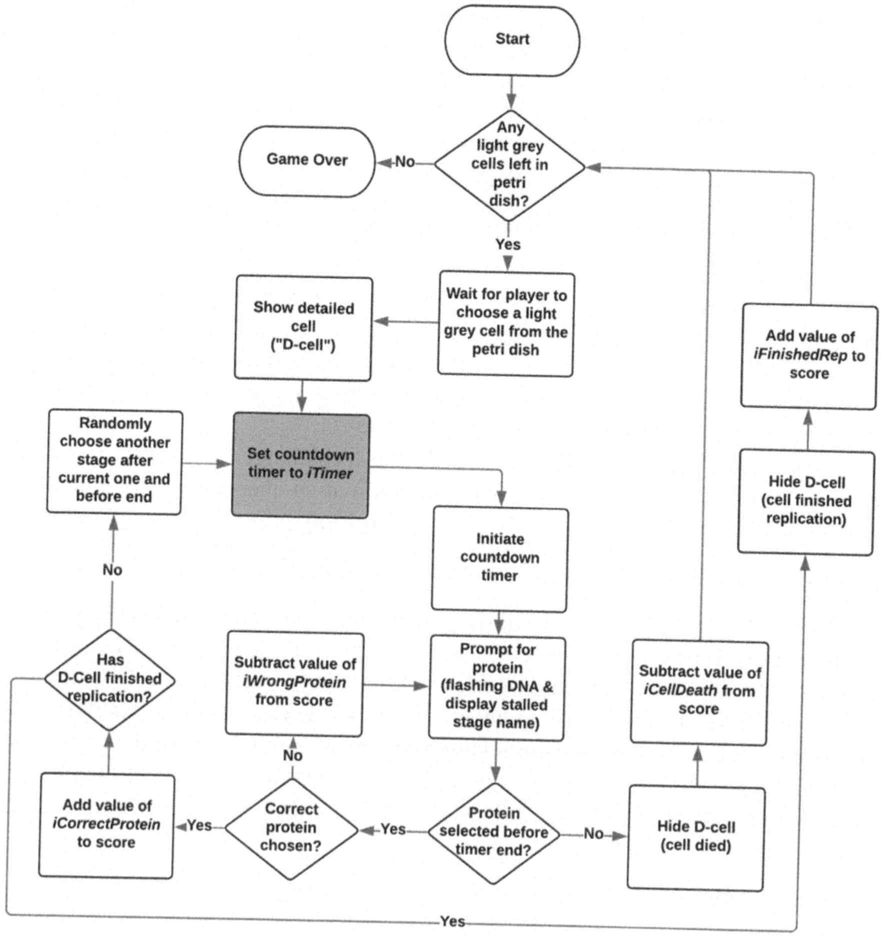

Fig. 6.6 Simplified logic flowchart for gameplay and scoring in *Okazaki's Revenge*

6.3 Reflecting on DVG Development

6.3.1 TCs' Reflections on DVG Development

The final assessment of the DVGs included a two-page reflection describing TCs' journey – from conceptualizing to playing the DVG. For the purpose of this chapter, three guiding questions that were part of the reflections were analyzed: *1) What do you feel were the most challenging aspects of developing a DVG? 2) What were your major successes in terms of developing a DVG? 3) What learnings do you feel you achieved that can be translated into your future practice?*

Reflections from 15 (seven females; eight males) out of 29 TCs' were randomly selected for the analysis. Participants' reflections were analyzed using NVivo12 software and constituted an interpretational analysis framework. Responses to the three questions were inputted as separate datasets into NVivo12 and analyzed for keyword frequency. Word clouds were generated through keyword frequency analysis, and this process was selected as it represented an aggregate of TC responses, as well as provided greater prominence to words (themes) that appear more frequently in the source text. The findings are illustrated below.

1. *What do you feel were the most challenging aspects of developing a DVG?*

TCs expressed feelings of being overwhelmed and intimidated by the assignment. The major barrier identified was a lack of programming experience, including coding, identifying an effective programming environment, and using variables (Fig. 6.7, *left*). In addition, TCs felt challenged in terms of their own learning process in researching the game subject matter, creating an engaging experience for students, and the frustrations of debugging their DVG.

Despite instructions to the contrary, scope was a challenge for many TCs as they initially storyboarded ambitious games approximating the complexity found in Nintendo's *Mario Kart* or Blizzard's *World of Warcraft*.

2. *What were your major successes in terms of developing a DVG?*

Overall, TCs were successful and recognized that their journey in DVG development improved their self-efficacy in terms of digital literacies (Fig. 6.7, *middle*). Many of the TCs expressed great pride in developing a functional DVG, despite their initial resistance to this assignment. The TCs observed that the experience of being out of their comfort zone during this assignment paralleled what their students will experience in practice. Furthermore, this assignment helped them internalize the importance and necessity for student support and scaffolding when implementing novel learning experiences in the classroom.

3. *What learnings do you feel you achieved that can be translated into your future practice?*

Fig. 6.7 *Left*: Word cloud highlighting TCs' challenges in DVG development. *Middle*: Word cloud highlighting TCs' successes in DVG development. *Right*: Word cloud highlighting TCs' learnings during DVG development

TCs recognized that success in DVG development was achieved through perseverance. In addition to learning STEM skills, such as coding and digital literacies, TCs also saw the possibilities for DVGs in their future practice (Fig. 6.7, *right*). Despite the fact that TCs felt the process of DVG development was valuable to them as practitioners, they perceived the exercise of creating a DVG as being more beneficial to their future students in terms of twenty-first century skills development and assessment. All but one of the 15 TCs' indicated that they intended to incorporate variations of the assignment in their future practice.

6.4 Reflecting on Implementation of DVGs in Practice

In the final reflection that was submitted with the completed DVGs, TCs were asked to reflect on their journey developing the DVGs, and whether they would incorporate DVGs in their future practice. All TCs were encouraged to play their peers' DVG and provide anonymous feedback. Below is a portion of one of the author's (a TC during the course) reflective account of implementing the DVG assignment with high school students.

In my third practicum, and after consultation with my associate teacher (AT), we decided to test DVG creation as an in-class assignment over three 75-minute periods with one grade 10 science (academic) and two grade 11 chemistry classes (university preparation). On the first day, students were introduced to *Scratch*, and were provided a worksheet that walked them through game design basics. By the end of the first class, most students (over 90%) had successfully created a Greek suffix-Roman numeral matching game. On the second day, the assignment was discussed: students had to create a *Scratch* game that focused on some aspect of their course. Similar to the response observed in my teacher education class, most students protested that this represented an unfair and unduly difficult assignment and appealed to the AT for relief. In response to student concerns, the AT chose to adopt the role of a student and complete the assignment alongside the students. By the end of the second day, most students had resigned themselves to creating a game, and were starting to enjoy the process. However, in each of the three classes, five to eight students appeared overwhelmed by the task. To support these students, on the third day I provided access to the *Scratch* prototypes I had previously created.

Many students chose to submit a variation on the Greek suffix-Roman numeral matching game; for example, matching chemical formulae and chemical name, or converting the matching game to a drag-drop game where chemicals had to be identified as ionic or covalent in nature. Most of the students that were still struggling at the end of the second day created *Jeopardy!*-style test preparation games that asked questions they themselves were struggling with, as study aides. Of particular note is the assignment submitted by a student with exceptionalities. George (pseudonym) was identified as being on the autistic spectrum, and had difficulty interacting with his classmates, being boisterous and highly opinionated with his peers. He struggled learning chemistry content but seemed to enjoy working independently. Of the 80+

students that completed this assignment, George was the only one that taught himself how to download a song from *YouTube* and implement an MP3 player within *Scratch*. His game used multiple backgrounds, changing color as more questions were answered correctly, and included a score counter. It was humbling to see a "problem student" excel to such an extent when properly challenged to succeed, and George represented a poignant reminder that all students are capable of far, far more than teachers believe.

All students were asked to complete a reflection as part of their assignment. Well over half the students felt that they learned something during the assignment and were interested in taking programming courses in the future. At my practicum school I was involved in the after-school Computer Club. After completing the assignment, seven students from these three classes began attending the Computer Club to learn more about programming. Of these seven, four were girls that were checking themselves back *into* the STEM pipeline, having previously identified as interested in the humanities or arts, and science only tangentially. Additionally, several girls during the three days of in-class work remarked to me that they didn't know they were smart; they thought only boys could program, and that it was a revelation to them that girls could be good programmers too!

6.5 Conclusion

This chapter highlights DVGs in STEM teacher education as an avenue for developing skills and practices essential for all students to learn, in accordance with the *Framework* (NRC, 2011). The authors argue that a variety of technological applications can be used to enhance STEM learning, such as digital timelines and video games, and urge the inclusion and modeling of content-specific pedagogical uses of technology in authentic contexts. In doing so, they provide theoretical and practical rationale for including DVGs, specifically for enhancing STEM literacy and STEM content knowledge, and developing STEM skills in teacher education.

In outlining the development of DVGs, in this case *Okazaki's Revenge*, the authors systematically demonstrate the potential of DVGs to address each of the STEM disciplines in a cohesive and comprehensive manner. The EDP is highlighted as a framework for effectively developing DVGs, as storyboarding and design documentation parallel the processes inherent in engineering design (Truesdell, 2014). Computational thinking, and accompanying skills, including the ability to resolve problems algorithmically and logically are also inherent in the DVG development. Furthermore, mathematics and science are embedded throughout *Okazaki's Revenge*; thus, conceptual understandings of science and mathematics are enhanced as players transfer their knowledge into new situations and apply it to new contexts, in this case DNA replication.

The creation of DVGs were instrumental in: i) providing opportunities for teachers to engage in the process of developing educative materials linked to curriculum, as exemplified in *Okazaki's Revenge*; ii) preparing them for using and modeling technology effectively in their future practices by enhancing and expanding their skill sets, such as programming, storyboarding and utilizing the EDP; and iii) engaging in best practices associated with integrated STEM education.

Particular tensions inherent in technology-focused activities, such as the DVG creation, are teachers' self-efficacy and beliefs about technology, as these are strong predictors of technological integration in practice. In order to increase TCs' self-efficacy and beliefs about the value of technology, as demonstrated in their reflections, they should be exposed to and experience ample opportunities to plan and select appropriate pedagogical practices in STEM. Moreover, TCs will need to understand the relationship between the affordances of a range of technological applications and detailed knowledge of STEM concepts, processes and skills (DeCoito & Richardson, 2016). This was highlighted through the TCs' acknowledgement of their initial lack of programming skills and their growth and maturation during their journey in conceptualizing and developing their DVGs.

Teacher education is at a new crossroads of incorporating technology effectively into pedagogical frameworks in STEM (Fullan, 2012). It is apparent that opportunities exist, such as DVG creation, that are catalytic for exploring new tools and processes to reinforce the alignment of technology and pedagogy in STEM teaching and learning. The authors contend that in order to achieve success in improving TCs' beliefs and self-efficacy related to technology, any strategy implemented should be explicitly modeled across a variety of disciplines in teacher education programs. This aspect was intentionally implemented in the DVG assignment. It will also require that teacher educators examine their STEM literacy, and accompanying content knowledge and skills, as well as their self-efficacy and beliefs about the value of technology as it applies to STEM education. This in turn necessitates identifying and establishing a vibrant pedagogical foundation for STEM teaching and learning.

Finally, curriculum will have to change to explicitly and systematically include technological integration, engineering and design thinking, as well as computational thinking and coding from K-12 education. If students are to engage in STEM learning by doing and be innovative and creative while using technology during the process, broader actions and discussions related to STEM education must occur with key stakeholders (e.g., government, policy makers, universities, school districts, business, teachers, parents, and students). DVG creation is just the 'tip of the iceberg' in the shifting landscape of opportunities for innovation in STEM education.

References

Abdalla, A., Forbes, N., Hall, J., Hipkiss, Hyde, P., O'Neill, M., et al. (2018). *Delivering STEM (science, technology, engineering and mathematics) skills for the economy*. London: Department for Business, Energy & Industrial Strategy and Department for Education, National Audit Office.

American Society for Engineering Education (ASEE). (2011). *"Engineering is..." engineering, go for it*. Washington, DC: Author.

Annetta, L. A. (2008). Video games in education: Why they should be used and how they are being used. *Theory Into Practice, 47*(3), 229–239. https://doi.org/10.1080/00405840802153940

Barata, G., Gama, S., Jorge, J., & Gonçalves, D. (2013). Improving participation and learning with gamification. In *Proceedings of the first international conference on Gameful design, research, and applications – Gamification* (pp. 10–17). Toronto, ON: ACM Press. https://doi.org/10.1145/2583008.2583010

Birk, M. V., Atkins, C., Bowey, J. T., & Mandryk, R. L. (2016). Fostering intrinsic motivation through avatar identification in digital games. In *Proceedings of the 2016 CHI conference on human factors in computing systems* (pp. 2982–2995). Santa Clara, CA: ACM Press. https://doi.org/10.1145/2858036.2858062

Bolstad, T., Gilbert, J., McDowall, S., Bull, A., Boyd, S., & Hipkins, R. (2012). *Supporting future oriented learning and teaching: A New Zealand perspective*. Wellington, NZ: Ministry of Education.

Bush, K., Gray, J., Holmes, M., Kosinski, K., Orr, J., Razzaq, L., & Rulfs, J. (2006, June). *How do you teach engineering in grades k and one?* (Paper presented at the Annual Conference & Exposition). Chicago, Illinois. Retrieved from http://peer.asee.org/702

Byl, P., & Hooper, J. (2013). Key attributes of engagement in a gamified learning environment. In *Proceedings of ASCILITE – Australian Society for Computers in Learning in Tertiary Education Annual Conference* (pp. 221–230). Retrieved from https://www.learntechlib.org/p/171232.

Conference Board of Canada. (2013). *Education and skills: Percentage of graduates in science, math, computer science, and engineering*. Ottawa, ON: Author. Retrieved from http://www.conferenceboard.ca/hcp/default.aspx

DeCoito, I. (2014). Teaching about the nature of science through digital scientific timelines. In P. D. Morrell & K. Popejoy (Eds.), *A few of our favorite things: Teaching ideas for K-12 science methods instructors*. Boston, MA: Sense Publishers.

DeCoito, I. (2016). STEM education in Canada: A knowledge synthesis. *Canadian Journal of Science, Mathematics and Technology Education, 16*(2), 114–128.

DeCoito, I. (2017). Addressing digital competencies, curriculum development, and instructional design in science teacher education. In *Encyclopedia of information science and technology* (pp. 1–12). Hershey, PA: IGI Global.

DeCoito, I., & Myszkal, P. (2018). Connecting science instruction and teachers' self-efficacy and beliefs in STEM education. *Journal of Science Teacher Education, 29*(6), 485–503.

DeCoito, I., & Richardson, T. (2016). Focusing on integrated STEM concepts in a digital game. In M. Urban & D. Falvo (Eds.), *Improving K-12 STEM education* (pp. 1–23). Hershey, PA: IGI Global.

Dicheva, D., Dichev, C., Agre, G., & Angelova, G. (2015). Gamification in education: A systematic mapping study. Educational Technology & Society, 18(3), 75+. Retrieved from https://link.galegroup.com/apps/doc/A427421595/AONE?u=googlescholar&sid=AONE&xid=4b95f3ae

Dym, C. L., & Little, P. (2000). *Engineering design: A project-based introduction*. New York: Wiley.

Fioriello, P. (2010). *Understanding the basics of STEM education*. Retrieved from http://drpfconsults.com/understandingthe-basics-of-stem-education/

Fullan, M. (2012). *Stratosphere: Integrating technology, pedagogy and change knowledge*. Toronto, ON: Pearson Canada.

Gray, L., Thomas, N., & Lewis, L. (2010). *Teachers' use of educational technology in U.S. public schools: 2009 (NCES 2010–040)*. Washington, DC: National Center for Education Statistics, Institute of Education Sciences, U.S. Department of Education.

Howard-Brown, B., & Martinez, D. (2012). *Engaging diverse learners through the provision of STEM education opportunities* (Briefing paper – South East Comprehensive Centre). US Department of Education.

Knippels, M.-C., Arend Jan Waarlo, P. J., & Boersma, K. (2005). Design criteria for learning and teaching genetics. *Journal of Biological Education, 39*(3), 108–112. https://doi.org/10.1080/00219266.2005.9655976

Koehler, M. J., & Mishra, P. (2009). What is technological pedagogical content knowledge? *Contemporary Issues in Technology & Teacher Education, 9*(1), 60–70.

Konstantinidis, A., Tsiatsos, T., & Pomportsis, A. (2009). Collaborative virtual learning environments: Design and evaluation. *Multimedia Tools and Applications, 44*(2), 279–304. https://doi.org/10.1007/s11042-009-0289-5

Lawson, M. (2019). *A first course in logic*. Boca Raton, FL: CRC Press.

Millar, R. (2006). Twenty first century science: Insights from the design and implementation of a scientific literacy approach in school science. *International Journal of Science Education, 28*(13), 1499–1521. https://doi.org/10.1080/09500690600718344

Mills Shaw, K. R., Van Horne, K., Zhang, H., & Boughman, J. (2008). Essay contest reveals misconceptions of high school students in genetics content. *Genetics, 178*(3), 1157–1168. https://doi.org/10.1534/genetics.107.084194

Mishagina, N. (2012). *The state of STEM labor markets in Canada*. Ottawa, Canada: Industry Canada.

NRC. (2011). *A framework for K-12 science education: Practices, crosscutting concepts, and core ideas*. Washington, DC: The National Academies Press.

Ontario Ministry of Education. (2008). *The Ontario curriculum grades 11 and 12: Science*. Retrieved from http://www.edu.gov.on.ca/eng/curriculum/secondary/2009science11_12.pdf

Orpwood, G., Schmidt, B., & Jun, H. (2012). *Competing in the 21st century skills race*. Ottawa, Canada: Canadian Council of Chief Executives.

Raven, P. H. (2007). *Biology*. New York: Glencoe McGraw-Hill.

Reddy, M., Bhaskara, V., & Mint, P. P. (2017). Impact of simulation-based education on biology student's academic achievement in DNA replication. *Journal of Education and Practice, 8*(15), 72–75.

Richardson, T., & DeCoito, I. (2017). *A case study of the impact of digital video games on students' physics career awareness* (Paper presented at the annual meeting of the Canadian Society for the Study of Education (CSSE)). Toronto, ON, May 27–31.

Roberts, C. (2010). *Introduction to mathematical proofs: A transition*. Boca Raton, FL: CRC Press.

Roehrig, G. H., Moore, T. J., Wang, H., & Park, M. S. (2012). Is adding the E enough?: Investigating the impact of K-12 engineering standards on the implementation of STEM integration. *School Science and Mathematics, 112*(1), 31–44. https://doi.org/10.1111/j.1949-8594.2011.00112.x

Ruggiero, D., & Mong, C. J. (2015). The teacher technology integration experience: Practice and reflection in the classroom. *Journal of Information Technology Education: Research, 14*, 161–178.

Schmidt, W.H. (2011). *STEM reform: Which way to go?* (Paper presented at the National Research Council Workshop on Successful STEM Education in K-12 Schools). Retrieved from http://sites.nationalacademies.org/cs/groups/dbassesite/documents/webpage/dbasse_072642.pdf

Stohlmann, M., Moore, T. J., & Roehrig, G. H. (2012). Considerations for teaching integrated STEM education. *Journal of Pre-College Engineering Education Research, 2*(1), 28–34.

Thakurta, R. (2013). Impact of scope creep on software project quality. *Vilakshan: The XIMB Journal of Management, 10*(1), 37–46.

Truesdell, P. (2014). *Engineering essentials for STEM instruction: How do I infuse real-world problem solving into science, technology, and math?* Arlington, VA\Alexandria, VA: NSTA Press\ASCD.

Vygotsky, L. S. (1978). *Mind in society: The development of higher psychological processes*. Cambridge, MA: Harvard University Press.

Wilson, S.M. (2011). *Effective STEM teacher preparation, induction, and professional development* (Paper presented at the National Research Council Workshop on Successful STEM Education in K-12 Schools). Retrieved from http://www7.nationalacademies.org/bose/STEM_Schools_Workshop_Paper_Wilson.pdf

Wing, J. (2014). *Computational thinking benefits society*. Retrieved from http://socialissues.cs.toronto.edu/index.html%3Fp=279.html

Isha DeCoito is an Associate Professor of STEM Education at Western University in Canada. Her research focuses on STEM engagement and STEM career aspirations amongst girls and underrepresented populations, gamification and other technologies in teaching and learning, engineering education, teacher professional development, and enhancing teacher candidates and students' Nature of Science conceptions.

Lisa K. Briona is the co-founder of Gamified Learning R&D. Her research focuses on engaging and retaining students by leveraging game mechanics in K-20 STEM education. She is the recipient of several education and educational technology awards recognizing innovation in technology enhanced teaching.

Chapter 7
A Self-Study on Teaching Integrated STEM Education to K-12 Science and Mathematics Teachers

Xinying Yin

7.1 Introduction

Quality Science, Technology, Engineering, and Mathematics (STEM) education is vital for the future success of students as well as global competitiveness for countries around the world. However, students are often disinterested in science and math when they learn in an isolated manner missing connections to crosscutting concepts and real-world applications (Kelley & Knowles, 2016). Integrated approaches to STEM education are gaining interests among educators and scholars. There is evidence that integrated STEM education can enhance student motivation, interest, achievement and persistence in STEM subjects, and in turn, these effects can help address the calls for greater workplace and college readiness skills and may increase the number of students who consider a career in a STEM-related field (Honey, Pearson, & Schweingruber, 2014). In the latest standards for K-12 mathematics and science in the U.S., both the Common Core State Standards for Mathematics (CCSSM; NGACPB, 2010) and the Next Generation Science Standards (NGSS; NGSS Lead the States, 2013) have called for more and deeper connections among the STEM subjects. The standards call for the engagement of students in authentic tasks that require integration across the STEM disciplines and emphasize the application of mathematics and science concepts. NGSS also provide specific guidance for improving K-12 education through the integration of computational thinking and engineering standards in science instruction.

Despite the increased attention to integrated STEM education, the STEM fields are in practice most often taught as isolated disciplines. Since integrated STEM education is a relatively new curricular approach, both teacher educators and K-12 teachers face challenges and ambiguities regarding the conceptualization and

X. Yin (✉)
California State University, Long Beach, CA, USA
e-mail: xyin@csusb.edu

V. L. Akerson, G. A. Buck (eds.), *Critical Questions in STEM Education*, Contemporary Trends and Issues in Science Education 51,
https://doi.org/10.1007/978-3-030-57646-2_7

pedagogy of integrated STEM education. Not only K-12 teachers, but also teacher educators who traditionally work in a single disciplinary field, often lack the understanding of and experience with integrated STEM education. In order to prepare teachers to teach integrated STEM, teacher educators must enhance their own understandings of integrated STEM education as well as how to facilitate K-12 teachers' development in this domain.

In this study, I, a science educator teaching a Masters' Program in STEM education, was developing and teaching an Integrated STEM Curriculum course for the first time. I have been a science educator for years, but the other disciplines of the "STEM" and the integrative ways of teaching STEM were new to me. Thus, driven by my desire to improve my understanding and teaching of integrated STEM curriculum and to facilitate K-12 teachers' development of integrated STEM teaching, I conducted a self-study in this course. This study is intended to advance the conversations around preparing K-12 teachers to teach integrated STEM education and enhancing STEM teacher educators' understanding and practice of teaching integrated STEM to K-12 teachers.

Research questions guiding this self-study are:

- How did the course activities facilitate (or not) students' learning of integrated STEM education?
- What have I learned about teaching integrated STEM education to K-12 teachers?

7.2 Literature Review

7.2.1 Defining Integrated STEM Education

Integrated STEM education can be generally viewed as the purposeful integration of the various disciplines as used in solving real-world problems and teaching the integrated disciplines in one cohesive entity (Breiner, Harkness, Johnson, & Koehler, 2012; Sanders, 2009). This is distinctively different from the traditional notion of STEM which often refers to the individual disciplinary coursework. However, ambiguities exist as what constitutes integrated STEM education because of the multiple ways such integration can occur. The integration may include different combinations of the STEM disciplines, connect a concept from one subject to a practice of another, combine practices from different disciplines, target one dominant subject with other supporting subjects; the integrated curriculum can also vary in the duration, setting, size and complexity, and involve a range of pedagogical strategies (Honey et al., 2014). A hallmark of integrated STEM education is the use of real-world situations or problems and integrated STEM experiences. It is suggested that a key mechanism of integrated STEM education is the cohesion of central concepts across the mathematics and science representations, engineering objects, design and construction activities, and social structures in the classroom. As well, it is important to design integrated experiences that provide intentional and explicit support for students to build knowledge and skills both within the disciplines and across disciplines (Honey et al., 2014). A more operational definition of integrated STEM education was proposed by Moore & Stohlman et al. (2014), as "an effort by educators to have students participate in engineering design as a means to develop technologies that require meaningful learning and application of mathematics and/or science" (p. 38). This definition of integrated STEM education specifically emphasizes the integration of engineering into science and mathematics classes.

In this chapter, integrated STEM education is defined as meaningful interdependence among all disciplines of STEM and this conceptualization of STEM was used to frame the course Integrated STEM Curriculum.

7.2.2 *Instructional Models for Teaching Integrated STEM*

Frameworks and instructional models of teaching integrated STEM have been proposed. These frameworks or models share commonalities in that they teach the STEM subjects in an integrated fashion by providing students with opportunities to apply STEM concepts and engage in STEM practices in interesting and relevant contexts. Moore & Glancy et al. (2014) identified a framework for integrating quality engineering education in K-12 science and mathematics classroom: (a) a motivating and engaging context, (b) an engineering design challenge, (c) opportunity to learn from failure through redesign, (d) the inclusion of math and/or science content, (e) student-centered pedagogies, and (f) an emphasis on teamwork and communication. This integrated STEM perspective introduces the use of an engineering design challenge to address a real-world problem. This framework was used in this study to guide students' lesson planning of an integrated STEM unit.

Further, several instructional models of teaching integrated STEM were introduced and discussed in the course. These STEM teaching models include context-based STEM education (Bybee, 2010), "T-SM-E" (Technology-Science & Mathematics-Engineering) model (Saito, Okumura, & Kumano, 2014), "6E Learning byDesignTM" model (Engage, Explore, Explain, Engineering, Enrich and Evaluate) (Burke, 2014), and project-based learning (Laboy-Rush, n.d.).

Bybee (2010) proposed a context-based approach to integrated STEM education. In this approach, students are introduced to a STEM-related challenge or problem they will confront as citizens, such as energy efficiency, climate change and hazard mitigation that is age-appropriate. When students explore the issues and gain an understanding of the problem, they must reach out to the STEM disciplines and apply knowledge and skills to address the issue, and the knowledge and skills would come from standards. Saito et al. (2014) described how they used a "T-SM-E" method to develop STEM lessons for a summer camp. The "T" is finding a technology in the society which may also include some problems to be solved, the "SM" is determining the science and mathematics content that need to be used to solve the problems of the technology, and the "E" is the engineering processes to solve the problems based on the "T-SM". Another instructional approach to integrating engineering and technology into science and mathematics instruction is using a "6E Learning byDesign™" model (Engage, Explore, Explain, Engineering, Enrich and Evaluate) built on the 5E instructional model (Bybee, 1997). In this model, students explore science and engineering concepts through inquiry-based instruction in the Exploration phase and apply the concepts to engineering solutions through development, construction, refinement, assessment, and redesign involved in the engineering processes. A project-based learning approach (Laboy-Rush, n.d.) consists of reflection (what is known and what needs to be investigated and learned to solve the problem), research (developing appropriate conceptual understanding of the project and its relevant concepts), discovery (bridging the research and information that is known with the project's requirements, initial solutions), application (model a solution, test the solution against requirements and possible revisions) and communication (a final stage where students present their solutions to peers and community).

The purpose of introducing the different integrated STEM teaching frameworks and models was to provide students with different means of conceptualizing the instructional approaches of integrated STEM so that they can have the flexibility and freedom to choose, combine and create their own approaches.

7.2.3 Developing Teachers' Expertise in Teaching Integrated STEM

K-12 teachers face various challenges of teaching integrated STEM, and the more generic approaches to teacher education has not been effective (Lederman & Lederman, 2013). In a study assessing teachers' professional development needs for integrated STEM education, the most salient finding was the universal belief that teacher education is wholly inadequate and need considerable improvements (Shernoff, Sinha, Bressler, & Ginsburg, 2017). Both internal and external barriers exist for teachers' implementation of integrated STEM. Some internal barriers include teachers' lack of exposure to other scientific and mathematical domains, little to no experience in engineering and technology skills, and limited familiarity to relevant instructional approaches to integrated STEM education(Asghar, Ellington, Rice, Johnson, & Prime, 2012; Park & Ertmer, 2008). The task of teaching integrated STEM itself is complex. It is challenging to attend to developing disciplinary tasks, the development of conceptual knowledge and reasoning, and application of knowledge in real-world contexts (Honey et al., 2014). When science teachers were implementing integrated STEM lessons, they also encountered challenges to make explicit connections between science, engineering, and mathematics while simultaneously maintaining a motivating and engaging context for their students throughout their instruction (Dare, Ellis, & Roehrig, 2018). STEM teachers also found that technology is the hardest discipline to integrate (Wang, Moore, Roehrig, & Park, 2011). External barriers include extended time needed to implement integrated STEM lessons, lack of adequate instructional materials, lack of effective assessments in integrated STEM, and lack of administrative support and encouragement (Asghar et al., 2012; Guzey, Moore, & Harwell, 2016; Moore & Stohlman, et al., 2014; Wang et al., 2011).

Since there is no one-size-fits-all for K-12 educators when it comes to an integrated STEM model (Bybee, 2013), teachers face the difficult task of determining what integrated STEM education means for them at a personal and practical level (Ring, Dare, Crotty, & Roehrig, 2017). In order to help teachers teach integrated STEM, their content knowledge base, as well as their understanding of how to integrate the STEM disciplines, must be increased (Honey et al., 2014). As well, teachers' conceptions of integrated STEM content influence how teachers conceptualize teaching integrated STEM. It was found that the degree of STEM integration that occurs in instruction may be related to teachers' ability to make explicit connections between the disciplines, and teachers' different conceptual models of integrated STEM lead to different ways of creating, developing, and writing integrated STEM curricula (Ring et al., 2017; Ring-Whalen, Dare, Roehrig, Titu, & Crotty, 2018). Thus, the content knowledge, the connections among the STEM disciplines, and the pedagogical approaches to integrated STEM education must be addressed. The Integrated STEM Curriculum course in this self-study was designed around these aspects.

Stohlmann, Moore, and Roehrig (2012) proposed that teachers need to develop student-centered pedagogy and self-efficacy with various support and necessary materials in order to successfully teach integrated STEM education. Several strategies to facilitate teachers' development of teaching integrated STEM include hands-on activities (Daugherty, 2009), teacher collaboration (Daugherty, 2009; Hunter, 2009; Slavit, Nelson, & Lesseig, 2016), and reflection on teachers' own conceptions of integrated STEM education both individually and with others (Ring et al., 2017). These strategies were employed in the course of this self-study.

7.3 Method

This inquiry was rooted in my desire to enhance K-12 science and mathematics teachers' understandings and practices of integrated STEM education. This study utilized a self-study approach. Self-study is a systematic, self-critical inquiry utilized by teacher educators as a way to not only improve practice but, more importantly, to come to a better understanding of that practice (Hamilton, 1998; Loughran, Hamilton, LaBoskey, & Russell, 2004). The self-study of teacher education practices is a powerful approach to promoting and sustaining change in teacher education (Bullough & Pinnegar, 2001; Cole & Knowles, 1998; Hamilton & Pinnegar, 1998). Teacher educators engaging in self-study aim to generate local, situated, provisional knowledge of teaching and to enhance understandings of teacher education practices, processes, programs, and contexts (LaBoskey, 2004). In this study, I, a science teacher educator, was teaching a new integrated STEM methods course in a

Master's program of STEM education for the first time. I explored how this course facilitated my students' understandings of integrated STEM education and how my own understandings of integrated STEM education evolved. Students' voices contributed to this study, and my understandings and practices were informed by students' learning and were evolving with the students'.

7.4 Context

This study took place in a course Integrated STEM Curriculum in a STEM Education Master's program in a medium-sized public university in Southwest U.S. At the time of the study, the program title was just changed from Mathematics and Science Education to STEM Education, intending to reflect the trend of integrating more engineering and technology in K-12 science and mathematics. The program courses included three science teaching methods courses (life science, physical science and earth-space science), three mathematics teaching methods courses (elementary mathematics, algebra and geometry), one inquiry in mathematics and science course, one assessment course, two research methods courses, one educational foundation course, and the Integrated STEM Curriculum course. The audience of the program was K-12 formal and informal educators who are interested in teaching the STEM subjects. The majority were elementary teachers and secondary mathematics and science teachers, and the rest were informal science educators and students who hadn't obtained a teaching credential yet.

The Integrated STEM Curriculum course was previously an integrated mathematics and science curriculum course. With the widely adopted NGSS standards that emphasize engineering and technology and the newly adopted program title "STEM Education", the program faculty decided to revise this course to an "Integrated STEM" course to reflect these changes. This course was the last in the program, designed to be a culminating course where students would make more connections between mathematics and science teaching. As transforming this course was the first step of revising the program curriculum, the course was the only course where engineering and technology was formally introduced to the students in this program.

The class met once a week for three and a half hours each time and lasted ten weeks. The overall goals of this course included developing students' understandings of engineering and technology and the interconnectedness among the four STEM disciplines, and developing their understandings and practices of integrated STEM education. As the students enrolled in this course in this study were mostly elementary and middle school teachers, the course activities were more cateredto them.

7.4.1 Course Activities

Instructional activities were adjusted with my ongoing analysis and reflection on students' learning. Personal reflections, class discussions and collaborative activities were employed. Major course activities included:

1. Use of video clips from the movie *Apollo 13* to elicit students' initial understandings of the STEM subjects and their interconnectedness.
2. Students' written reflections and class discussion on a series of readings of integrated STEM education, including both the theoretical foundations and the different models of teaching integrated STEM. The readings were from different resources such as *A Framework for K–12 Science Education: Practices, Crosscutting Concepts, and Core Ideas* (National Research Council, [NRC], 2012), *STEM Integration in K-12 Education: Status, Prospects, and an Agenda for Research* (Honey et al., 2014) and various peer-reviewed journal articles on integrated STEM education.
3. Frequent students' written reflections on their conceptions of STEM and teaching integrated STEM throughout the course.
4. "Engineering a story" activity (Dubosarsky, 2014) for students to experience and understand the engineering design process and then develop a lesson that integrates science and math standards based on their engineering design activity.
5. Visiting a local STEM middle school and reflective discussions on integrated STEM education.
6. Collaboratively creating a context-based integrated STEM unit.

More details of the course activities can be found in *Appendix 1*. Weekly Course Activities.

7.4.2 Participants

The participants of the study included myself and seven students in this course who were K-12 science and mathematics teachers. I also invited a mathematics educator to some of my planning sessions and a few class sessions so she can help me with her expertise in mathematics education. I did not collaborate with her in teaching this course or conducting this self-study because of the limited common time we had and administrative restrictions on co-teaching a course. At the time of the study, I was a junior faculty just started teaching in the Master's program and had several years of experience teaching elementary and secondary science methods before. I was also a former high school chemistry teacher. Out of the seven students, five were females and two were males. There were four elementary teachers, one middle school mathematics teacher, a pre-service elementary teacher and a pre-service secondary science teacher. The five in-service teachers' teaching experience ranged from 1 year to 5 years.

7.4.3 Data Collection

Data resources for this self-study included transcripts of selected class discussions during two audiotaped class meetings, students' artifacts including their written reading reflections, in-class reflections and lesson plans developed in this course, pre- and post- surveys about their understandings of STEM disciplines as well as teaching integrated STEM, and my lesson plans and reflective journals. Two class meetings, one on the discussion about the school visit, and another on students' reflecting on their final STEM unit, were videotaped. Students' written reflections focused on their understandings developed from reading materials and class sessions, how their conceptions of integrated STEM education was evolving, and the questions they had. These reflections, along with other assignments, provided ongoing feedback to me and informed my instructional decisions.

Pre-survey questions included: (1) What is engineering? Please give an example with your explanations. (2) What is technology? Please give an example with your explanations. (3) What do you think are the relationships among science, technology, engineering and mathematics? – You can use a concept map along with some descriptions/explanations to illustrate your ideas. (4) What do you think is meant by integrated STEM education? Provide an example of an integrated STEM instructional lesson/activity. (5) What do you think are the advantages of teaching integrated STEM curricula? (6) What do you think might be the barriers or difficulties of teaching integrated STEM curricula? Post survey had the same six questions and one more question about students' comfort level and confidence of teaching an integrated STEM unit: How would you describe your ability to construct and implement a STEM unit or lesson? In other words, how confident and comfortable are you with designing and implementing a STEM unit or lesson?

My personal reflective journal was guided by the following questions:

- To what extent are students' understandings and practices about STEM education developing?
- What do I need to do in the next two weeks? What should I do to meet students' needs or respond to their questions?
- Whether, and how, are my understandings about STEM education evolving?

7.4.4 Data Analysis

Data analysis was an iterative process occurring concurrently with data collection and at the completion of the study. All students' reflections were analyzed ongoingly and provided me with opportunities to evaluate the efficacy of current course activities and make adjustments in the subsequent classes. I also periodically discussed emergent themes and future directions for practice with the mathematics educator being consulted for this course. This continuous analysis became part of the data, which, in turn, served as additional data for analysis at the completion of the study.

Data analysis took two stages, direct interpretation and categorical aggregation (Stake, 1995), which was an inductive process of narrowing from the particular codes to larger themes (Creswell, 2007). Each data resource, including transcripts of selected class discussions from videotaped class meetings, students' written reading reflections, in-class written reflections, lesson plans developed by students, pre- and post- surveys, and my lesson plans and reflective journals were analyzed and then synthesized to answer the research questions. Episodes of class discussions were selected when they were focused on students' conceptions of integrated STEM education. Since the class size was small, all reading reflections, in-class written reflections, students' lesson plans, and surveys were included in the data analysis. The analysis process was validated by peer review and repeated readings of the original data. Along with data analysis during the project, member-check was done with students to validate data and the researcher's interpretations (Creswell & Miller, 2000). Peer-debriefing was done with the mathematics educator in the final phase of data analysis. Findings in the final report were then reviewed for completeness and credibility (Guba, 1978).

7.5 Findings

1. How did the course activities facilitate (or not) students' learning of integrated STEM education?

Throughout the course, reading reflections and discussions, hands-on activities, visiting a STEM middle school, creating STEM lessons and units, and constant reflections on the course activities collectively helped students develop their conceptions about integrated STEM education.

Various readings about integrated STEM education were provided to help students build conceptual foundations about engineering and technology, as well as integrated STEM education. The reading reflections were guided by specific prompts that helped students to make connections to their conceptions and practices of teaching integrated STEM. Students also shared and discussed their reading reflections in class to clarify their questions and strengthen their understandings. For example, students reflected on how their ideas about engineering and technology have evolved three weeks into the course:

> I never thought of engineering on its own as a key concept that students must understand. After starting this class and the readings for the week, I realized that engineering is very important and is equally important in the STEM process of learning for students. (student written reflection)

> Before week 1 of classes, I thought of technology as something people used to communicate and convey information like computers, software and electrical apparatuses. I know

now that technology is much more; it can be anything that assist people in performing a task. (student written reflection)

Students also raised significant questions along with their reflections that guided my planning. For example, one question was "How can I effectively help the students come up with a question or issue that is most important to them that can be incorporated into a project that we can do as a class?". In responding to this question, I planned an activity where students discussed the criteria of a good driving question or issue for an integrated STEM unit before delving into planning the unit.

The visit to a local STEM school generated valuable discussions about integrated STEM curriculum. Students discussed how the teachers were integrating STEM or not. For example, for a Lego Robotic class, they discussed that there was a possible lack of content taught in terms of the technology because the students were just following procedures and going through trial and error without necessarily understanding how the robotics worked. They discussed how an ecology class which was working on a water quality project was a better example of integrated STEM because the teacher explicitly talked about the chemistry concepts and prepared students for an engineering project of purifying water. They also discussed how to integrate more mathematics into a city planning project. Observing real classrooms provided students with real-world pictures of how integrated STEM might look like, and these discussions demonstrated students' developing understandings about integrated STEM education.

Particularly, the lesson planning activities that emphasized integrating content standards from the STEM disciplines facilitated students to develop their personal working strategies to plan integrated STEM lessons and their conceptions about integrated STEM education. About half-way through the course, students were asked to consider a "personal working strategy" to designing an integrated STEM lesson/unit after completing the "Engineering a story" activity (Dubosarsky, 2014) where students did a hands-on engineering activity, reading articles on the different approaches of integrated STEM teaching and trying to integrate science and mathematics into their engineering activity. Many of them addressed the importance of standards and learning objectives, finding a real-world problem that could relate to those standards and relevant to students, and making sure the content is taught explicitly and not in a superficial way. For example, one student commented:

> I would start with the learning goal in mind, then identify the societal problems that relate to the learning goal. The next step would be to incorporate the standards for all STEM components and combine them in a way that allows for a seamless transition from one area to the next. (student written reflection)

Some of the students also mentioned using project-based or the "6E" model to design their integrated STEM lesson, doing research to enhance their knowledge in the different content areas, and seeking input and feedback from colleagues.

Another critical activity in this course was a three-week-long project where all the students collaboratively developed a unit around the topic of carbon dioxide (CO_2) levels in mobile homes for upper elementary students. This activity started with discussing and determining the criteria for a "good" issue or driving question for an integrated STEM unit. The criteria students came up with included: it is relevant to students, students are able to design experiments or products and obtain results to answer the question, it allows multiple approaches, the question/issue can lead to learning in different disciplines from grade-level standards, and materials are available to the teacher and students. Then they went through iterations of deciding on an issue and the content standards that could be developed into a meaningful integrated STEM unit. They worked in small groups to plan the different lessons focusing on specific subject areas in this unit. After initial lesson planning, they presented and peer-critiqued the initial lesson plans, and went through several rounds of discussion on how to revise and sequence these lessons so that they were connected coherently. They then revised their lesson plans, did final presentations and reflected on the process of creating an integrated STEM unit. The unit included a science activity on learning about CO_2 and the respiration process producing CO_2, measuring CO_2 levels at home at different locations throughout the day, graphing the results and interpreting the graph, understanding the unit ppm and calculating volumes in a room related to CO_2 quantities, building and modifying a CO_2 detector, researching how to reduce CO_2 levels at home and finally connecting the issue to CO_2 levels globally.

Students contributed their different expertise to this collaborative activity, as well as experienced the frustrations and challenges of constructing an integrated STEM unit. This experience helped them develop clearer understandings of the integrated STEM curriculum, recognize the importance of collaboration in developing integrated STEM unit, "uncover multiple layers of challenges" (class discussion), "have a starting point to implement integrated STEM" (class discussion) and feel more confident in creating STEM lesson plans. They pointed out several important elements for an integrated STEM lesson or unit, including allowing for inquiry, the issue and concepts being relevant to students and being able to integrate different disciplines, having multiple approaches to solving a problem, knowledge- and standard-based, grade-level appropriate and feasible to implement. Content standards became prominent in considering an integrated STEM unit. A student commented, "My initial understanding was that I would create a unit and then assign standards that are relevant to the lesson. But when looking at the unit from the outcome backward, or reverse planning as it's called in educational articles, it becomes more difficult" (student written reflection).

Collaboration is one most salient aspect that students mentioned when reflecting on the process of creating this integrated STEM unit. They appreciated the opportunities to work with others and provide feedback to each other. They also expressed the challenges they saw in creating the integrated STEM unit, such as being unfamiliar with standards outside of their disciplinary area and finding a topic that is relevant to a wide diversity of students. Although some questions remained, such as the grade level appropriateness, rigorousness of the content, and assessment issues, this experience was overall very positive. Several students indicated that they were thinking how to start implementing integrated STEM in their classrooms and they felt more confident of doing so.

At the end of the course, students developed substantial new conceptions about integrated STEM education. These included their understandings about engineering, technology and the relationships between the STEM disciplines, their conceptions about teaching integrated STEM curriculum and their perceptions of the advantages and challenges of teaching integrated STEM.

First, students' understandings about engineering, technology and the relationships between the STEM disciplines became more sophisticated. The students started this course with some narrow understandings about engineering and technology, such as engineering as constructing buildings and technology as electronic devices. At the end of this course, all of the students explained that engineering was the design, creation or improvement of human-made products that improve people's life and/or a process of solving problems, consistent with "Engineering is a systematic and often iterative approach to designing objects, processes, and systems to meet human needs and wants" (National Assessment Governing Board, 2010, as cited in NRC, 2012, p.202). While only five examples were given by students on the pre-survey, there were 15 examples given on the post-survey, such as designing different pencils for different hands and creating better methods of recycling. Students' conceptions about technology also expanded to be more consistent with the definition used by *A framework for K–12 science education* (NRC, 2012, p.202): "Technology is any modification of the natural world made to fulfill human needs or desires" (National Assessment Governing Board, 2010). For example, one student's initial response was "technology is the use of electronic-based gadgets"(pre-survey), and it changed to "technology ranges from soles on our shoes to satellites in space" (post-survey).

With their developed conceptions about engineering and technology, students also expanded their views about the relationships among the four STEM disciplines. In the beginning, all students recognized that the four STEM disciplines were definitely related, but the relationship was described in general and simplistic terms such as "they are closely related to each other" and "engineering and technology are the application of science and mathematics" (pre-survey). At the end of the course, the relationships were described in more sophisticated ways and the interdependence among the four disciplines were emphasized. The following response shows that this student viewed the four STEM disciplines as intertwined and propelling each other to advance in a "never-ending cycle":

> Through that science we collect data, and that data is analyzed through the mathematics skills and practices we learn. The understanding of the science and mathematics to understand the worldly ideas we are curious about then makes us wonder about how we can make things easier on us, thus transitions to technology. The technology can always be improved upon and that is where the engineering comes in and helps us build new technology, and better modify the old technology to make it more relevant in our lives. The relationships intertwine as new improvements and ideas are always coming about and further scientific concepts and mathematics skills are explored to enhance the technology through the engineering processes. It is a never-ending cycle and relationship among the four important parts to learn and grow with. (post-survey)

These developed conceptions of engineering, technology and the relationships among the STEM disciplines helped students build a foundation in thinking about teaching integrated STEM curriculum. Although they all mentioned that integrated STEM curriculum was the integration of the STEM subjects at the beginning, they could not provide any details of the integration and their definitions of integrated STEM did not necessarily include engineering. At the end of the course, all students indicated that an integrated STEM curriculum should include engineering and technology integrated with mathematics and science, sometimes also with social studies and English language arts. They also emphasized that the different subject areas should be integrated in a relevant and coherent way. Further, they became intentional and purposeful in considering the content to be integrated. They provided specific examples with explicit elaborations about how the different STEM subjects were embodied in a lesson/unit. For example, a student described a lesson that attempted to "find a solution to improve the means of transportation" in a city she lived, because "currently transportation is scarce and citizens of the town are having difficulty entering and leaving the town to run personal errands and/or attend higher educational schools." She explained:

> Science: The science portion would be researching and reading about how vital it is for a town such as … to have regularly scheduled bus routes entering and leaving the town.
>
> Technology: Students can use technology to speak to people in charge of scheduling bus routes in order to make it more accessible to its citizens.
>
> Engineering: Students can devise a better means of transportation for the town, such as a van carpool or devising a better bus.
>
> Mathematics: Students may perform mathematical economical reasons of why transportation is beneficial in such a town. (post-survey)

When students were discussing their examples, some pointed out that the "entry point" of designing an integrated STEM lesson/unit can be flexible: "I think that all the concepts are connected to each other in many ways. A teacher can start with any of the subjects and still cover the rest" (post-survey). As well, they also mentioned that creating a STEM lesson can start with a contextual issue or start with curriculum standards. This indicated that they were forming their personal approaches to integrated STEM lessons. In the students' reflections and post-surveys, they mentioned their own preferred model such as the "6E Learning byDesign™" model (Burke, 2014), and project-based learning model (Laboy-Rush, n.d.), which were more operational rather than conceptual.

Besides, students' perceptions of the advantages and challenges of teaching integrated STEM curriculum expanded through the course. Students recognized some advantages of teaching integrated STEM curricula at the beginning, including more relevant, engaging and meaningful real-world situations, and the possibility to improve students' critical and creative thinking, problem-solving skills, communication, and deeper conceptual understanding. At the end of the course, they provided more elaborated rationales and put more emphasis on teaching standards, helping students make connections among these disciplines and students having different "entry points" to learn. One student commented, "students get more out of the lesson especially those who believe that they are weak in a particular area. The spoonful of sugar (science) offers students to invest in the portions of the unit or lesson that contain math" (post-survey). Students also pointed out the benefits of teaching integrated STEM for teachers, which was not mentioned before nor was anticipated. They said that teaching integrated STEM can enhance teacher's engagement because of the inquiry-based approaches and creativity involved, enhance time efficiency of teaching multiple subjects in one unit, and more importantly, provide the opportunity to teach more science in elementary schools.

Students' perceptions about the challenges of teaching integrated STEM also changed through the course. In the beginning, students had more concerns about how to teach integrated STEM curricula, such as lack of subject matter knowledge, lack of pedagogical content knowledge, and lack of examples. At the end of the course, some students mentioned the challenges to ensure the integrated content at appropriate grade level and rigor, as well as the difficulty in assessing the multiple concepts in an integrated manner. They also expressed concerns about administration, the time needed for lesson planning, cost, resources, and teacher collaboration opportunities. These concerns indicated the increased awareness of the conditions for implementing integrated STEM curricula. Only by first recognizing these constraints can teachers start thinking about how to overcome them.

2. What have I learned about teaching integrated STEM education to K-12 teachers?

Although I had studied a collection of literature on integrated STEM education and had some experience with engineering activities at conference workshops and presentations, I started teaching this course with many uncertainties. My understandings were mostly on paper and I had no experience teaching engineering or integrated STEM in a K-12 classroom, not to mention teaching others how to teach integrated STEM. I had uncertainties about what integrated STEM meant for K-12 classroom, what an integrated STEM unit would exactly look like, and what the process of developing an integrated lesson or unit would entail. During this course of the self-study, my understanding of integrated STEM education was evolving with the students'. Many things that my students learned contributed to my own learning. I was learning with my students, and I consider we collectively developed more understandings about integrated STEM education in this course.

First, the different integrated STEM instructional models were very helpful in guiding the teachers to construct an integrated STEM lesson. When teachers understand the different models, they can choose one model or combine different models flexibly to realize their instructional goals. In this course, students developed their first integrated STEM lesson where they first came up with a "technology" from a children's story (Dubosarsky, 2014), engineered the technology and then integrated relevant science and mathematics content to plan a lesson. Here we discussed the "T-SM-E" (Technology-Science and Mathematics- Engineering) model (Saito et al., 2014) as a conceptual model to think about the content of this lesson and using project-based learning model as a delivery method. When designing the integrated STEM unit on CO_2 level in mobile homes, we started with brainstorming meaningful issues that K-12 students may face as discussed in the context-based STEM education proposed by Bybee (2010) and then the students chose to use the "6E Learning byDesign™" model (Burke, 2014) to design this unit.

In creating the integrated STEM unit and reflecting on the process, the students and I realized that there were multiple "entry points" of designing an integrated STEM. The lesson planning can start with selecting an issue/context or selecting content standards. In working on the final collaborative STEM unit, students went through several iterations of determining the topic and matching grade-level appropriate content standards. As well, the initial point of integration could occur at any discipline. A student in this course with a technician background tended to think of a technology (e.g., CO_2 detector, a body lotion) as the starting point of her integrated STEM unit; and a student with a science education background tended to consider a science-related topic (e.g., environmental issues) as the starting point.

Through this course, the students and I also came to better understandings about the complexities and challenges of developing an integrated STEM unit. These included: a meaningful and engaging context in which students can explore important content from different subject areas, the coherence and connections of different lessons/activities to support the overall goal of the unit, assessment of students' learning of STEM integration, and contextual factors in implementing an integrated STEM unit/activity such as constraints of time, resources and students' experience. If teachers' content knowledge is not strong enough, the integration will be more difficult. As well, teacher collaboration is very important but also challenging. In collaboration, teachers need to understand the goals of the whole unit and each other's lesson, so that they can collectively develop lessons that can meaningfully tie together under a certain context/driving question. Such collaboration also requires a high level of collegiality and peer-support because different ideas need to be united coherently. In this course, the students worked together well because they have known each other for a long time, their collaboration was structured and purposefully facilitated, and sufficient time and space were provided.

Last but not least, I have realized how difficult it would be to integrate all the four disciplines of STEM when attempting to teach integrated STEM curriculum. The conceptualization of integrated STEM education as meaningful interdependence among all disciplines of STEM was adopted for the course. The lesson planning activities in this course asked students to integrate both mathematics and science standards with engineering and technology. These lesson planning activities were all done in collaborative groups where students had different expertise, so it seemed to work fairly well. In reflection, trying to integrate all four disciplines at the same time can be daunting and complicated, especially for teachers beginning to integrate STEM, because the teachers have to be familiar with all subject areas and find logical connections among these subject areas. Further, when a teacher does want to integrate both mathematics and science content into an integrated STEM unit, as some of the students mentioned, the weight of each subject area is often not equal and depending on the teacher's expertise and the nature of the topic. In my subsequent year of teaching this course, I started the lesson planning activities to require engineering integrated with mathematics and/or science, which allowed students to integrated mathematics or science with engineering, and this seemed to help ease students' anxiety and transition from single subject teaching to integrated STEM teaching.

7.6 Discussions and Implications

7.6.1 Successes, Limitations, and Improvement of this Course

Overall, the students' understandings about engineering, technology and the relationships between the STEM disciplines became more sophisticated, and they have developed substantial understandings about teaching integrated STEM through this course. Their understandings about engineering and technology became more consistent with how they are defined in the public documents and literature such as *A Framework for K-12 Science Education* (NRC, 2012) and *STEM Integration in K-12 Education: Status, Prospects, and an Agenda for Research* (Honey et al., 2014). Their views of the relationships among the four disciplines in STEM also became more multi-dimensional and interdependent. Their developed conceptions of engineering, technology and the relationships among the STEM disciplines helped them conceptualize teaching integrated STEM curriculum. This was consistent with the previous research findings that teachers' conceptions of integrated STEM content influence how teachers conceptualize teaching integrated STEM (Ring et al., 2017; Ring-Whalen et al., 2018).

With regard to teaching integrated STEM, we only focused on the planning stage in this course due to time limits. Through reading literature, collaborative activities, lesson planning, class discussions, and constant reflections, students have developed their working definitions of teaching integrated STEM in K-12 classrooms. Visiting the STEM middle school provided opportunities for students to visualize what integrated STEM would possibly look like in K-12 classrooms and generated valuable discussions on how they perceived the STEM lessons they observed. The three-week-long collaborative effort on creating an integrated STEM unit, where students experienced the real challenges as well as successes of planning integrated STEM lessons/unit, was crucial to facilitate students' practice of teaching integrated STEM.

Bybee (2013) suggests that it is important that teachers consider their own conceptions of integrated STEM at the local level. At the end of the course, all of the students indicated that integrated STEM curriculum included engineering and technology integrated coherently with science and mathematics; and sometimes other subject areas like social studies and English language arts can also be integrated. They came to see the importance of standard and learning objectives, being intentional and purposeful in integrating content standards, finding a real-world problem that could relate to those standards and relevant to students, making the content explicit, and teaching the different content areas in a cohesive manner. They were also forming their personal approaches to plan an integrated STEM lesson by using different teaching models.

Due to my limited expertise and experiences of integrated STEM education, limited resources I had, limited time of the course and no real teaching opportunities, there were limitations in students' learning of teaching integrated STEM. First, engineering and technology were only discussed at a general level. Many details were not discussed in depth, such as the constraints and resource, and the optimization processes in engineering. Further, when students were describing examples of integrated STEM lesson in the post-survey, the roles of technology and mathematics were more of supporting rather than prominent learning objectives. Although "technology" is inherently included when students "engineer" something, technology is not explicitly taught or discussed. This might be due to my lack of expertise in mathematics and technology thus less explicit discussions and reflections on these two subject areas. Previous studies also found that technology was considered by teachers to be the hardest discipline to integrate (Wang et al., 2011) and mathematics and technology tend to be missing from STEM educational approaches (Herschbach, 2011; Rinke, Gladstone-Brown, Kinlaw, & Cappiello, 2016). Although students have developed some personal working knowledge of planning integrated STEM, there were some challenges that the course did not adequately address, including ensuring all the content areas are integrated at the appropriate level and rigor, and the difficulty of assessment in integrated STEM lessons while maintaining an engaging learning context for students (Dare et al., 2018; Honey et al., 2014).

Regarding the integration of the subject areas in integrated STEM, teaching the four areas of STEM cohesively within an authentic context may sound ideal, but it has been found to be a difficult task for teachers (English, 2016; Rinke et al., 2016). In this course, the students collaboratively developed their integrated STEM unit that included all four disciplines. However, I saw students' struggles of making the four disciplines fit together into one lesson/unit. It would have been easier for the students to learn to plan integrated STEM lessons had I asked them to start with integrating engineering with mathematics or science but not both.

In the future, I may consider inviting outside experts on mathematics, engineering and technology as guest speakers, starting from simpler integrations instead of all four subjects, analyzing and critiquing field-tested examples of integrated STEM lessons, and having more explicit discussions on the role of technology and mathematics.

7.7 K-12 Teachers' and Teacher Educators' Learning to Teach Integrated STEM Education

Integrated STEM education does not only benefit students' learning but is also a promising approach to encourage teachers to improve teaching science and mathematics. In this self-study, I saw that learning to teach integrated STEM stimulated teachers to change how they teach, since integrated STEM "forced" them to use real-world contexts and inquiry-based teaching strategies. Another important benefit to teachers was that it could help elementary teachers to more efficiently teach multiple subjects and have more opportunities to teach science under the pressure of focusing on math and English language arts in the current accountability system.

However, as trying anything new in one's teaching, learning to teach integrated STEM takes a lot of courage, tolerance of uncertainties, and a stance of learning with students. I was modeling the process for my students in teaching this course. Embarking on the journey of teaching integrated STEM to K-12 teachers was a bold move for me. I was very open to my students that this was a new experience for me, and we were learning together. I felt uneasy throughout the course when students kept asking me for good examples of integrated STEM lessons, but I was not sure what a good example looked like. However, when I was looking at students' reflections and post-surveys, I was somewhat relieved that they did learn many new things about integrated STEM education and they felt this was a successful experience.

I found myself as a facilitator and collaborator of students' learning, and we were exploring integrated STEM education together in this course. When working on the lesson plans and units, I did not know what the lessons or the units would turn out to be. I was mainly guiding the processes with some principles of integrated STEM education, such as meaningful context, explicit instruction on standard-based content, and the coherence among the different components of the lesson or unit. My lack of expertise and experience in the course content was undoubtedly undesirable. However, using a "guided collaborative" approach empowered both me and my students to develop our working knowledge of integrated STEM education. This process also modeled for my students how teachers needed to be willing to and be comfortable with taking risks and learning with students when they were trying something new. A previous study on teachers learning to teach integrated STEM also found that risk-taking is important in teacher development (Slavit et al., 2016).

Although I have learned a lot about integrated STEM education through this self-study, I wish I could have some systematic professional development. A challenge of professional development on integrated STEM education was the time needed. This Integrated STEM Curriculum course was built on the foundations of previous inquiry-based science and mathematics teaching methods courses. Also, this course took a total of ten weeks and 40 instructional hours. It was still far from enough for the students to develop a solid understanding of engineering, technology and the various issues involved in teaching integrated STEM. As a course in a Master's program, one solution is to have another course focusing on engineering and technology before students come to the integrated STEM curriculum course and modify other science and mathematics methods courses to include STEM integration. For K-12 teachers' professional development, this means that a long-term, systematic program.

The challenges of collaborating with other teachers exist in both higher education institutions and K-12 schools. In teaching this course, I consulted a mathematics teacher educator about the mathematics component in the integrated STEM, but I did not find other faculty who had expertise in engineering education in my institution. In the future, I may use virtual networks to seek colleagues from different disciplines who have similar goals in teaching integrated STEM education. The limited collaboration with the mathematics teacher educator in this study resulted in mathematics being a weaker component in this course. In order to more fully collaborate with other faculty members, co-teaching could be a better way to do so. However, this could involve administrative barriers, as well as a significant amount of time devoted by all the parties to have productive discussions, come to understand each other's discipline, and negotiate different ideas. Similarly, K-12 teachers need to find like-minded people to start the collaboration and seek support from schools and districts for the time and spaces needed for long-term collaboration.

If integrated STEM education is to be effectively implemented, both K-12 teachers and teacher educators need substantial professional development and support to develop the expertise needed in new areas of the STEM disciplines as well as the pedagogy of teaching integrated STEM. This study is situated in the United States, but the issues faced by K-12 teachers and teacher educators in teaching integrated STEM could share commonalities. I hope this self-study provides an opportunity of conversation and learning for teacher educators from different countries who aspire to advance integrated STEM education.

Appendix 1: Weekly Course Activities

Week	Activity
1	Pre-survey about students' conceptions of STEM discipline and STEM education. ***STEM Disciplines in the real world***-- *Where do you see science, engineering, mathematics and technology* *In Apollo 13 movie clips?* Summary and reflection: (1) compare and contrast the four disciplines S.T.M.E (2) collectively illustrate the relationships among the disciplines.

(contimued)

Week	Activity
2	***Explore "what is engineering and technology"?*** 1. Reading discussion. 2. Personal written reflection: Reflecting on how your ideas about engineering and technology have 3. Evolved so far. Leave any questions you want to further discuss/explore about. ***An introduction to STEM integration framework*** Honey, M. Pearson, G. & Schweingruber, H. (2014). STEM integration in K-12 education: Status, prospects, and an agenda for research. Committee on Integrated STEM education; National Academy of engineering; National Research Council. Washington, DC: The National Academies Press.
3	***STEM instructional models*** - Reading discussion. ***"Engineering a story" activity***. Workshop presented at Association for Science Teacher Education Conference, San Antonio, TX (2014, January)
4	***"Engineering a story" activity continued & Design a STEM lesson plan*** Students brought their materials and constructed what they designed last time for solving the problem they identified in the "little Bear's trousers" story. 1. Reflection and discussion -Use the Informed Design Rubric (Crismond & Robin, 2012) to evaluate their own practices Crismond, David P., & Adams, Robin S. (2012). The informed design teaching and learning matrix. *Journal of Engineering Education, 101*(4), pp.738–79 2. Create an integrated STEM lesson based on their activity
5	***Design a STEM lesson plan continued.*** Class reflection: What are the important elements in a "successful/effective" STEM lesson/unit?
6	Student presented their reading reflections of seven articles or book chapters. **<u>After class reflection:</u>** Based on what you heard and discussed in class, and using people's reflections/ presentations in week 5, summarize three pieces of the most important information (or new ideas) regarding STEM teaching you have learned from the 7 article presentations/ activities. Please devote at least one well-written paragraph for each idea (e.g. one topic sentences supported by 3–4 sentences and one conclusion sentence).
7	***Visit a local STEM school*** Reflection and discussion for what is or is not integrated "STEM" education.
8	***Designing a context-based integrated STEM unit as a group.*** Pre-project activity: What makes a "good" issue/driving question What are the standards you can address during the unit Students divided the "unit" into different parts and lessons to work in small groups.
9	***Designing a context-based integrated STEM unit as a group*** Students brought in their individual lesson plan and presented it. Both the instructors and peers gave feedback. Student took it back and revised their lesson plans.
10	***Designing a context-based integrated STEM unit as a group*** Students presented their revised lesson plans. We reflected on the STEM unit and the designing process. Post-survey about students' conceptions of STEM discipline and STEM education.

References

Asghar, A., Ellington, R., Rice, E., Johnson, F., & Prime, G. M. (2012). Supporting STEM education in secondary science contexts. *The Interdisciplinary Journal of Problem-based Learning, 6*(2), 85–125. https://doi.org/10.7771/1541-5015.1349

Breiner, J. M., Harkness, S. S., Johnson, C. C., & Koehler, C. M. (2012). What is STEM? A discussion about conceptions of STEM in education and partnerships. *School Science and Mathematics, 112*(1), 3–11. https://doi.org/10.1111/j.1949-8594.2011.00109.x

Bullough, R. V., & Pinnegar, S. (2001). Guidelines for quality autobiographical forms of self-study research. *Educational Researcher, 30*(3), 13–21.

Burke, B. N. (2014). 6E learning byDesignTM model: Maximizing informed design and inquiry in the integrative STEM classroom. *Technology and Engineering Teacher, 73*(6), 14–19.

Bybee, R. W. (1997). *Achieving scientific literacy: From purposes to practices*. Portsmouth, NH: Heinemann.

Bybee, R. W. (2013). *A case for STEM education*. Arlington, VA: NSTA Press.

Cole, A. L., & Knowles, J. G. (1998). The self-study of teacher education practices and the reform of teacher education. In M. L. Hamilton (Ed.), *Reconceptualizing teaching practice: Self-study in teacher education* (pp. 224–234). London: Falmer Press.

Creswell, J. W. (2007). *Qualitative inquiry and research design: Choosing among five approaches* (2nd ed.). Thousand Oaks, CA: Sage.

Creswell, J., & Miller, D. (2000). Determining validity in qualitative inquiry. *Theory Into Practice, 39*, 124–130.

Dare, E. A., Ellis, J. A., & Roehrig, G. H. (2018). Understanding science teachers' implementations of integrated STEM curricular units through a phenomenological multiple case study. *International Journal of STEM Education, 5*, 4. https://doi.org/10.1186/s40594-018-0101-z

Daugherty, J. (2009). Engineering professional development design for secondary school teachers: A multiple case study. *Journal of Technology Education, 21*(1), 10–24. https://doi.org/10.21061/jte.v21i1.a.1

Dubosarsky, M. (2014, January). *Engineering a story* (Workshop presented at Association for Science Teacher Education Conference), San Antonio, TX.

English, L. D. (2016). STEM education K–12: Perspectives on integration. *International Journal of STEM Education, 3, 3*. https://doi.org/10.1186/s40594-016-0036-1

Guba, E. (1978). *Toward a methodology of naturalistic inquiry in education evaluation* (CSE monograph series in evaluation no. 8). Los Angeles: University of California, Center for the Study of Evaluation.

Guzey, S. S., Moore, T. J., & Harwell, M. (2016). Building up STEM: An analysis of teacher-developed engineering design-based STEM integration curricular materials. *Journal of Pre-College Engineering Education Research, 6*(1), 1–19. https://doi.org/10.7771/2157-9288.1129

Hamilton, M. L. (Ed.). (1998). *Reconceptualizing teaching practice: Self-study in teacher education*. London: Falmer Press.

Hamilton, M. L., & Pinnegar, S. (1998). Introduction: Reconceptualizing teaching practices. In M. L. Hamilton (Ed.), *Reconceptualizing teaching practice: Self-study in teacher education* (pp. 1–4). London: Falmer Press.

Herschbach, D. R. (2011). The STEM initiative: Constraints and challenges. *Journal of STEM Teacher Education, 48*(1), 96–112.

Honey, M., Pearson, G., & Schweingruber, H. (2014). *STEM integration in K-12 education: Status, prospects, and an agenda for research* (Committee on integrated STEM education; National Academy of engineering; National Research Council). Washington, DC: The National Academies Press.

Hunter, M. S. (2009). *Morriss math and engineering elementary school: A case study of K–5 STEM education program development*. Columbus, OH: The Past Foundation.

Kelley, T. R., & Knowles, J. G. (2016). A conceptual framework for integrated STEM education. *International Journal of STEM Education, 3*, 11. https://doi.org/10.1186/s40594-016-0046-z

LaBoskey, V. K. (2004). Afterword. Moving the methodology of self-study research and practice forward: Challenges and opportunities. In J. J. Loughran, M. L. Hamilton, V. K. LaBoskey, & T. Russell (Eds.), *International handbook of self-study of teaching and teacher education practices* (Vol. 12, pp. 1169–1184). Dordrecht, The Netherlands: Springer. https://doi.org/10.1007/978-1-4020-6545-3_29

Laboy-Rush, D. (n.d.). *Integrated STEM education through project-based learning*. Retrieved from http://www.rondout.k12.ny.us/common/pages/DisplayFile.aspx?itemId=16466975

Lederman, N. G., & Lederman, J. S. (2013). Is it STEM or "S & M" that we truly love? *Journal of Science Teacher Education, 24*(8), 1237–1241. https://doi.org/10.1007/s10972-013-9370-z

Loughran, J., Hamilton, M. L., LaBoskey, V. K., & Russell, T. L. (Eds.). (2004). *International handbook of self-study on teaching and teacher education* (Vol. 12). Dordrecht, The Netherlands: Springer.

Moore, T. J., Glancy, A. W., Tank, K. M., Kersten, J. A., & Smith, K. A. (2014). A framework for quality K-12 engineering education: Research and development. *Journal of Pre-college Engineering Education Research, 4*(1), 1–13. https://doi.org/10.7771/2157-9288.1069

Moore, T. J., Stohlman, M. S., Wang, H. H., Tank, K. M., Glancy, A. W., & Roehrig, G. H. (2014). Implementation and integration of engineering in K-12 STEM education. In S. Purzer, J. Strobel, & M. Cardella (Eds.), *Engineering in pre-college settings: Synthesizing research, policy and practices* (pp. 35–60). West Lafayette, IN: Purdue University Press.

National Assessment Governing Board. (2010). *Technology and Engineering Literacy Framework for the 2014 National Assessment of Educational Progress*. Retrieved from http://www.nagb.org/publications/frameworks/prepub_naep_tel_framework_2014.pdf

National Governors Association Center for Best Practices (NGACPB). (2010). *Common Core State Standards for Mathematics*. Retrieved from www.corestandards.org/assets/CCSSI_Math%20 Standards.pdf

National Research Council. (2012). *A framework for K–12 science education: Practices, crosscutting concepts, and core ideas*. Washington: National Academies Press.

NGSS Lead States. (2013). *Next generation science standards: For states, by states*. Washington, DC: The National Academies Press.

Park, S. H., & Ertmer, P. A. (2008). Examining barriers in technology-enhanced problem-based learning: Using a performance support systems approach. *British Journal of Educational Technology, 39*(4), 631–643.

Ring, E. A., Dare, E. A., Crotty, E. A., & Roehrig, G. H. (2017). The evolution of teacher conceptions of STEM education throughout an intensive professional development experience. *Journal of Science Teacher Education, 28*(5), 444–467. https://doi.org/10.1080/1046560X.2017.1356671

Ring-Whalen, E., Dare, E., Roehrig, G., Titu, P., & Crotty, E. (2018). From conception to curricula: The role of science, technology, engineering, and mathematics in integrated STEM units. *International Journal of Education in Mathematics, Science and Technology, 6*(4), 343–362. https://doi.org/10.18404/ijemst.440338

Rinke, C. R., Gladstone-Brown, W., Kinlaw, C. R., & Cappiello, J. (2016). Characterizing STEM teacher education: Affordances and constraints of explicit STEM preparation for elementary teachers. *School Science and Mathematics, 116*(6), 300–310. https://doi.org/10.1111/ssm.12185

Saito, T., Okumura, J., & Kumano, Y. (2014, January). *The STEM education research in Japan and its prospective* (Paper presented at the Association for Science Teacher Education Conference). San Antonio, TX.

Sanders, M. (2009). STEM, STEM education, STEM mania. *The Technology Teacher, 68*(4), 20–26.

Shernoff, D. J., Sinha, S., Bressler, D. M., & Ginsburg, L. (2017). Assessing teacher education and professional development needs for the implementation of integrated approaches to STEM education. *International Journal of STEM Education, 4*, 13. https://doi.org/10.1186/s40594-017-0068-1

Slavit, D., Nelson, T. H., & Lesseig, K. (2016). The teachers' role in developing, opening, and nurturing an inclusive STEM-focused school. *International Journal of STEM Education, 3*, 7. https://doi.org/10.1186/s40594-016-0040-5

Stake, R. (1995). *The art of case study research*. Thousand Oaks, CA: Sage.

Stohlmann, M., Moore, T. J., & Roehrig, G. H. (2012). Considerations for teaching integrated STEM education. *Journal of Pre-College Engineering Education Research, 2*(1), 28–34. https://doi.org/10.5703/1288284314653

Wang, H., Moore, T. J., Roehrig, G. H., & Park, M. S. (2011). STEM integration: Teacher perceptions and practice. *Journal of Pre-College Engineering Education Research, 1*(2), 1–13. https://doi.org/10.5703/1288284314636

Xinying Yin is an Associate Professor of Science Education at California State University- San Bernardino, and a former high school chemistry teacher. She teaches preservice science teachers and a Master's program in STEM Education. Her research interests include preservice and inservice science teachers' conceptions and practices of integrated STEM Education, formative assessment and culturally relevant pedagogy.

Chapter 8
Learning for the Real World: Interdisciplinary Challenge Projects to Facilitate Real-World Learning in STEM

Jessica Dobrin

8.1 Project-Based Learning

While project-based learning is increasingly popular as a means of providing authentic learning experiences, there is much disagreement on what qualifies as project-based learning and indeed what qualifies as an authentic learning experience. There exists such a diverse body of literature on project-based learning (Hasni, Bousadra, Belletête, Benabdallah, & Dumais, 2016; Thomas, 2000) that to declare decisively for one definition or another would prove problematic at best, and it is not my intention to do so here. Rather, the point of this work is to discuss one means of making PBL authentic. While this requires a coherent definition for both authentic learning and project-based learning, it is not my assertion that these definitions are definitive. That being said, for the sake of clarity, I will provide a brief introduction to the definitions I am working from, so as to avoid any confusion.

It may reasonably be assumed that in project-based learning, regardless of any other particulars, a project of some type must be involved. Such a statement appears initially to be so simplistic as to imply circular logic, but it is dependent on a coherent definition of a project. Kokotsaki, Menzies, and Wiggins (2016) suggest that the project is the creation of a tangible artefact, with prior learning being utilised and new learning or new knowledge developed as necessary for the successful completion of the assignment. There seems to be general agreement that the project cannot be peripheral to the curriculum (Bender, 2012; Thomas, 2000), though precisely how central the project must be, and indeed what is considered curricular are more fluid criteria. In the UTC model (more below) the project itself is a component of the school-wide curriculum, with different disciplines as the focus of the project throughout the year.

J. Dobrin (✉)
Dulwich College, Beijing, China
e-mail: jessica.dobrin@cantab.net

V. L. Akerson, G. A. Buck (eds.), *Critical Questions in STEM Education*, Contemporary Trends and Issues in Science Education 51,
https://doi.org/10.1007/978-3-030-57646-2_8

Once a project has been established, it may be reasonably assumed that the aim is that, through this project, some type of learning is achieved. This is I believe a more functional definition that it might initially appear; it is certainly possible for situations to occur in which learning is complemented by a project, or for a project to be a means of displaying learning that has already taken place. These would not, in my view, constitute project-based learning. This is not to say that all learning must take place exclusively through the project either. There seems to be general consensus between practitioners and researchers (Bender, 2012; Mills & Treagust, 2003; Prince & Felder, 2006) that project-based learning focusses more on the application rather than the 'acquisition' of knowledge, though none deny the possibility that some new knowledge might be constructed. Indeed, it may be considered unlikely that any project be completely devoid of opportunities for new knowledge or understanding to be developed, and so new learning of this type must also be considered when discussing the aims of a PBL project.

To focus exclusively on the content knowledge applied during the project would also be a disservice to the PBL model; learning encompasses more than just content knowledge. Skills such as communication, teamwork, planning, problem-solving, analysis, and evaluation may be developed during the PBL project (Dobrin, 2020). These skills, alternatively referred to as 'soft', 'transferrable' (Canelas, Hill, and Novicki, 2017; Carvalho, 2016), or even 'twenty-first century', form critical components of significant learning (Fink, 2013) outcomes for students. Indeed, it might be argued that one of the primary benefits of PBL over other teaching models is that the focus on application of knowledge allows for the development of these skills within a curricular context, which may aide in the transfer of knowledge from the classroom to the real world. If we take these components together, we can define PBL as *learning through an authentic project, central to the curriculum, through which students apply or develop knowledge that may be transferred to the real world.*

The transfer of skills from the classroom to the real world is a component of authentic learning as well. Indeed, much of the literature treats 'real-world' and 'authentic' as synonymous terms; Herrington and Oliver's (2000) framework for authentic learning provides a prime example of this in their treatment of elements of situated learning. Thomas (2000) posits that an essential characteristic of the PBL project is that it is 'realistic', or provides a feeling of authenticity. Kokotsaki et al. (2016) make reference to authentic problems and questions while defining PBL. A manufactured scenario with no connection to the real world cannot result in a true PBL project. When considered alongside the focus on application rather than acquisition of knowledge, the PBL project is a means through which students can apply existing content knowledge and in doing so develop skills that can be utilised in the real world. These are the transferrable, twenty-first century skills considered above. This authenticity relies not on content area, but rather the structure and context of the problem or task around which the project is based.

While it is true that authenticity is not limited to a particular discipline, there has been a concentrated push create direct transference of skills from classroom to career in STEM (science, technology, engineering, and mathematics) subjects. STEM is considered by some to be a meta-discipline (Brown, Brown, Reardon, &

Merrill, 2011) greater than the sum of its parts; if this is so, too will the skills necessary for success in the field have to go beyond numeracy, scientific literacy, or thinking like an engineer. Indeed, some argue that one of the defining features of STEM education is the integration of its constituent parts into work that parallels that of a scientist or an engineer (Breiner, Harkness, Johnson, & Koehler, 2012). STEM-based PBL projects may offer students the opportunity to develop these skills in a realistic setting, whether the project is conducted within the syllabus of a single subject, an integrated STEM classroom, or as an addition beyond the core subject areas. In the following sections, I will consider two different PBL projects: a hypothetical example based in a standard high school physics classroom, and an example from a STEM-based school in the UK in collaboration with local engineers. I will then discuss the projects in terms of their outcomes, and lay out the instructional implications of each.

8.2 Methodology

While the first, hypothetical, example is not the result of a particular school study but rather an accumulation of personal experiences over the years, the second example, the Challenge project, is the result of targeted research undertaken as part of my doctoral degree. The description of the Challenge project is the result of pre-project meetings with the instructional team, as well as weekly observations for the duration of the project. The opinions of the students are the products of a series of interviews conducted throughout the year. One student, Student A, was a participant in the hydropower dam project that forms the primary focus. Student A was at the time of the project enrolled in the first year of a two-year GCSE[1] Physics course, which would cover forces, energy, eaves, electricity, magnetism, particulate model of matter, atomic structure, and space. The other students, Students B, C, and D, were older students in the same school, referred to here as Riverside College.[2] These interviews, conducted one or twice per week for each student, offer valuable insight into student experience within the Challenge project. Students from a second school, referred to here as Parkside College, provided additional input through targeted one-off focus groups, but because of the structure of their projects did not participate as regular contributors in the same manner as Students A-D.

The entirety of the research, like PBL itself, was constructivist in nature, meaning that I as the researcher was constructing meaning from the collected data, and it is this meaning I am attempting to convey in this chapter. The primary research aim had been to make a study of the measurable learning outcomes of students engaged

[1] GCSE is a course of study conducted over 2 years, when students are approximately 14–16 years of age, US grades 9–10. Typically, students study Biology, Chemistry, and Physics concurrently over this period.

[2] College is used according to the UK tradition, referring to secondary rather than undergraduate education

in these Challenge projects, through the means of semi-structured interviews; that data is primarily presented elsewhere (Dobrin, 2020). While outcomes are certainly addressed within this chapter, the focus is instead on the experience of the individual participants, and their opinions on what works within these projects, as well as what does not. Any attempt to entirely separate outcomes and experience would be to do a disservice not only to the participants, but to the PBL instructional model designed to put students at the centre of their learning. All narratives relating to the Challenge projects will therefore represent a mix of outcomes and experiences, providing a student-centred perspective on a unique STEM-driven PBL model.

8.3 A Hypothetical Example

A teacher in a standard high school Physics classroom decides that having students design a hydropower dam model is the best way to allow students to apply several key curricular points regarding energy, motion, and conservation. Because the students do not have an extensive engineering background, the teacher provides specifications for a model of a hydropower dam, which the students will be able to construct using provided supplies. In this way, students are focussed less on the practical design, and more on the targeted concepts meant to be reinforced by the project. Students are placed into groups for this project to minimise the supplies needed, and to help them develop transferrable skills relating to communication and teamwork. In order to increase student autonomy, the teacher provides basic design instructions, but leaves it to the students to determine some specifics such as flow rate and net head. Students use a hydropower equation to make their predictions for power output based on values they are provided or could directly measure from their models. Students have already had introductory lessons relating to electricity and power, so much of this is review for them, though the context is new. Depending on the style of the teacher, the lessons may be delivered as the project progresses, or direct instruction may be front-loaded in an introductory session to allow students to self-regulate later in the project.

In total, 6 days are dedicated to this project. The first lesson introduces the project, establishes the groups, and depending on how the project is structured, students may be provided brief lessons on the power equation, flow rates, energy demands, and other relevant information. Students may then be given a short amount of time in their groups to plan their next steps. Once students take some time, either in class or as homework, to determine the specifications of their dams, they come up with a list of materials necessary for their models, so that the teacher can have enough supplies for the following day.

The next three lessons are dedicated to the building of the hydropower dam models, and a fifth day is dedicated to testing the models, before then hooking up a multimeter and determining actual power output. This then leads to an interesting discussion-how close are the students' predicted values, and why do they differ? A discussion about efficiency ensues, and one group asks if they can modify their original design to improve this. They come in after school that day, and make some small changes to the turbine which produce a moderate change, and so the next day in class they start the lesson by sharing what they've learned with their peers. Each group then goes away and writes up a lab report, to be submitted for assessment. In some schools this report is an individual effort, while for others a single group report is submitted.

8.4 The Challenge Project

In the hypothetical example above, the instructor is limited primarily by the school's timetable, and the number of lessons that can possibly be allocated for each topic. While individual schools may have allowances for co-curricular activities or even unstructured time dedicated to enrichment or further study, the notion of dedicating significant blocks of time to project-based learning is simply impossible in many school settings. Even the most well-designed projects are therefore often fragmented, with individual sessions of perhaps an hour (or less) being pieced together to meet the project aims. For this reason, the teacher may encourage students to get together outside the school day to have longer uninterrupted periods to work. If the aim of instruction is to encourage students to apply content knowledge in authentic circumstances and develop transferrable skills, such a model is still preferable over a lecture-based unit, though some limitations remain.

In a second example, I draw on an actual project conducted at a STEM-focussed University Technical College (UTC) in the East of England, Riverside College. The project is fit into a system referred to by the school as Challenge, wherein 1 day per week, regular lessons are suspended in favour of a long-term project, often if not always supported by a local employer in a science or engineering field. Challenge projects are conducted outside the core curricular subjects, often based in one particular STEM discipline, but designed in such a way as to require skills from across STEM to successfully complete each Challenge. The project observed for instance draws not only on a main discipline of Physics, but builds in Maths, Biology, Geology and Geography, and Engineering principles and skills as well. Objectives are designed in such a way as to inextricably link these disciplines in the spirit of STEM education; the engineering elements require the use of Maths and Physics, and decision-making often requires an understanding of local ecology or geology as well.

This particular project is, like the previous example, undertaken with the goal of creating a hydropower dam, and is being led by members of a local engineering firm, in collaboration with an internal instructor. The dam being designed for this Challenge was in fact built by this company, and so in that regard this example differs from the previous one in that there is an established "right" answer to the student work, though students are not immediately made aware of this. The motivation for the project is also based more firmly in the final product than in the application of content knowledge from a particular course, as the main facilitators are not responsible for covering specific information from the syllabus with these projects. Rather, a collaboration between the school and the employers establishes an appropriate skill level (as suggested by the school) and target skill areas (suggested by the employers) necessary to complete the dam model.

The structure of this project also differs; because there are seven full days (1 day per week for 7 weeks) allocated for this project, there is more time to establish some context. On the first day (Week One), students spend the morning getting a brief introduction to the project objectives. Then, students are asked to consider the need for more sustainable energy production, calculate their own energy usage, and brainstorm ways they could reduce this. An activity involving designing an energy plan for the UK results in some hilarity, as students choose deliberately outrageous solutions before settling in to consider how to best balance energy needs against available resources. This activity led students to then consider the energy needs of the nation of Georgia, where the dam is to be located. This first Challenge day ends with a short recap, where students are encouraged to pull together their ideas.

The session in the second week of the project sees students first introduced to the structure of a working hydropower dam, as well as the equation used for calculating power. A brief lesson on topography and map reading leads to students trying to select a best possible location for their dam and turbine. The students are allowed three possible locations, and they then consider the pros and cons of each. Sites are rejected for having too low a flow rate, being too near a town, or having a topographical profile unsuitable for the type of dam they had selected. Students get the chance to explore this further in the afternoon when they are asked to brainstorm solutions to such potential problems as disrupting local tourism, destroying the habitat of endangered species, or destroying areas of historic or cultural value. Again the solutions range from flippant to extremely thoughtful; one group has taken the time to research local customs to improve relations with the locals, and their ideas are received with great enthusiasm by their peers and facilitators alike. Students who have less interest or experience with these tasks begin designing an informational website, which will be used to display the findings of each team, as well as detailed models and descriptions of the dams once completed.

The third week is split between geotechnical engineering and using computer software to create a 3D model of the dam. Those interested in geotechnical engineering remain in the main Challenge area to learn about how soil composition, topography, and building materials may influence the stability of their dams. The students are given data relating to standard penetration test results, which they then graph and use to select a single value to represent the average for their own dams. Students are encouraged to develop arguments based on their data to justify their choices. Some of the students spend the afternoon session writing up their findings from this portion, while the remainder focus on the 3D modelling task. Some students choose to import the topographical features of their chosen location to add authenticity to the project, while others opt for a more generic model. Those who had been less interested in the calculations in the morning become more interested in learning the features of the software, and some even create additional designs.

The fourth week is focussed on the output of hydropower dams. The morning features presentations of the types of energy involved in hydropower dams, types of dams, and how energy is converted into useable forms. This leads to a discussion about the selection of a type of turbine. Returning to calculations completed in the second week of the project, the groups use their previously collected or calculated data relating to power output to select turbines. Students also consider efficiency of different turbines based on the nature of the dam location (net head, flow rate, etc.). The groups make their selections, and a small model made of popsicle sticks is attempted, but not completed due to time constraints. While this model little serves to reinforce any of the calculations made earlier, it allows the students a chance to experience manipulating materials, rather than focus entirely on the abstract. At this point, the majority of the decisions regarding the group's dam have been made, and the final details are put on the website they have been designing to advertise their dam's specifications.

Week five is a bit of a stand-alone lesson, in that the students are now considering a different aspect of their dams, the structural integrity of the walkway. One student recalls that a dam burst could flood the area and cost both lives and money, and so the importance is immediately seen. A morning session related to theory and generic calculations leads to the creation of models made primarily of chocolate, which will be tested in the final week of the project. Students are given a set 'budget' and cost list for additional materials ranging from barbecue skewers to oat cereal, and attempt to balance out the strength of their materials against a limited amount of funds. Teams differ wildly in their strategy, with some risking pure chocolate for their models, others meticulously planning as accurate a model as possible with the strongest available materials, while others opt for as much of the cheapest materials possible hoping that quantity will win out over quality.

Exams in the week following the building of the 'chocrete' beams allow them an extra week to harden before students gather for week six, where the models are tested using buckets of water to test the maximum weight they can hold. The testing results in much hilarity as the students attempt to break their chocolate beams, and allows the students to finish the practical portion of the project on a high note. Students are also given a presentation on the real dam project, and given the chance to compare their dams to the one built by the engineers who guided them through the project. Several of the issues they considered are also revealed to have come up during the construction of the actual dam, and at least one group is pleased to discover their solutions matched the ones used out in the 'real world'. Students are also offered feedback on their websites, as well as their dam designs, with some groups then asked to present their projects at a college open day later in the term.

8.5 The UTC Concept

The Challenge project is one of several employer-facilitated project models being used at University Technical Colleges (UTC's) across the UK. A relatively recent (2009) addition to the UK education system, UTC's were conceived as a means of addressing a shortage in skilled labour in industry, particularly science and engineering (McCrone, White, Kettlewell, Sims, and Rush, 2019). Graduates attending university chose STEM-related subjects at a rate of 74% (compared to a national average of 46%), and UTC leavers gained apprenticeships at rates nearly four times the national average (University Technical Colleges (UTC), 2019), suggesting that the model is effective in meeting its goals. UTCs each have a primary focus area, often in STEM, but UTCs are not exclusively STEM training institutions. Rather, the UTC model is based in the belief that participation from the local community, particularly local employers and universities, is critical for preparing the students for the next stage of their learning. Students choose to enter a UTC because they have a commitment to the focus of that particular school, and are seeking out learning *for* the real world they will enter upon completion of their programmes.

Industry involvement in these schools may take many different forms, depending on the school focus. The two UTCs I observed form a perfect illustration of this. Riverside College in the example above has a broad science and technology focus, with an aim to introducing students to the biotech or engineering industries prevalent in the area. Due to an increasing number of local employers with a focus in computer sciences, these are also common projects as well. Students in years 10 and 11 complete a number of Challenge projects that over the course of a year will be based in a number of different main subject areas, while sixth form students are given the chance to select from a range of projects each cycle according to their interests. In all cases, projects are approximately 9 weeks in duration and have one full day per week dedicated to their completion. Depending on local interest, these projects are designed either in close collaboration with a local employer, or internally by staff with experts in the field making guest appearances. Many of the projects also involve site visits to research facilities, local waterways, etc.

A second UTC, 'Parkside College', located in the South East of England, has a primarily engineering focus due to demand in the area. The school benefits from several partnerships with nearby universities and employers, notably automotive engineering and manufacturing in the area. As such, there is a strong emphasis on engineering, though students generally consider themselves to be either an "engineering student" or a "science student" and approach the projects accordingly. Both science and engineering are represented in the projects run, here called Employer Led Projects. Unlike the previous UTC, students here are assigned their projects into sixth form as well, though care is given to ensure a variety of disciplines and interests are catered to. Less time is dedicated to projects here, only a few hours per week on a given morning, but because in most cases all students within a year group participate in the same project, there is more potential for teachers to reference these projects during regularly scheduled class time. Many of the projects at this UTC contain a competition element, where students work in teams to win prizes or industry experiences. Panels of judges from the employer-partner mean that students receive feedback not just on their presentation, but on how what they have done fits within the industry itself. One project for example was judged by a panel of industry experts in engineering, including an apprentice who had entered the field through his own exposure to the industry as part of the same programme years before. This afforded the students feedback from an individual closely acquainted with both the student and the industry perspectives, which several students reported finding valuable. This and other student experiences and perspectives are presented below.

8.6 Key Findings

The following sections outline many of the key findings of the research, primarily features of the PBL project, their impact on the students, and the outcomes that result. The first of these is autonomy and student agency, and the impact these have on the students. Next, I consider the authenticity of the projects, and their applicability both to the core curriculum and to the students' futures. Finally, I will consider the impact of real-world experts on the projects, both in terms of student motivation and career networking. While Student A forms the primary focus of this report, Students B, C, and D also offered insights from their own projects that prove useful here, either as support for Student A's perspective or as an equally valuable contrast.

8.7 Student Agency and Autonomy

Among the most critical characteristics of PBL, as a constructivist teaching method, is the agency of the students. Students treated as active agents in their own learning make choices, with guidance, about how to best reach their goals. Student A's Challenge project, the hydropower dam project, was explicitly and thoughtfully designed to incorporate an element of student agency: most sessions featured two or more different tasks suited to the interests and abilities of the different learners, something Student A was quick to notice and appreciate:

> Student A: I think we had a lot of flexibility to do what we wanted and the fact that we had the freedom to do what we wanted and the way we could do this on our own was great yeah

Student A felt that even within the project criteria provided to him, he and his group were able to do things in a way that made sense to them. That he and each of his teammates were able to have such differing roles was a frequent topic of discussion, nearly always in a positive light. Student A felt he was making positive contributions to his group because he was able to utilize his strengths. This same sentiment was found in the Computer Science Challenge project, with one participant sharing the following insight:

> Student B: …all of them do Computer Science A Level and I was quite like I guess I was quite good at doing like the report and the Power Point and everything and it all just worked out we all just kind of fell in to our own places in the end anyway like I don't know it was just nice yeah

Student B was placed outside her comfort zone for her project but, due to the nature of her assignment, she was able to play a direct and valuable role in her group's performance. The importance of her position as a non-programmer in a project based on programming was reinforced by her teammates, her instructors, and the professional mentors who came in to support them. This was due in large part to the team's understanding that in the industry, such differentiation is common, and the task was such that the marketing and presentation carried as much weight as the final code. Not all teams took such a differentiated approach; in this case it was possible due to the backgrounds and skills of the team members.

What was quite important in these projects was the deliberate design of challenges that allowed for this student-driven differentiation to take place. In the case of the hydropower dam project, part of the purposeful overhaul from the previous year was the move from a paper report and poster session to an informative website, including a 3-dimensional digital model. It was believed that this would allow students who did not feel comfortable in the engineering aspects of the project to work in building a website according to their abilities. In some groups, students chose to rotate through the roles depending on the engineering task, while others opted to stick to differentiated roles throughout. Likewise, in each case it was intended that the students determine this for themselves, with one instructor remarking that "you know that project learning is working because I don't have anything to do".

In terms of outcomes, this student agency very much allowed students to develop skills in problem-solving, planning, inner-group communication, even argumentation. Students found it necessary to consider the tasks before them, math the tasks up with the abilities of one or more teammates, set a course of action for themselves, and then justify the choices that were made. Student A took on the role of Lead Engineer for his group because he had a great deal of interest in Engineering and Maths, so he found the idea of calculating energy output for his dam to be enjoyable. His teammates with more interest in the design elements were happy to take his calculations and use them to complete the digital design of the team's hypothetical dam. Not all groups chose such a differentiated approach; while Hamish and his team split the project into Engineering tasks, Maths tasks, and Design tasks, other teams worked more collaboratively on each task.

Not all projects feature the same level of agency, however. Student C, involved in a Water Management Challenge project at approximately the same time, was acutely aware of the lack of agency in his project:

> Student C: well I mean hand-holding there's too much of it there's too much not there's not enough free choice provided for the students and so I feel like you can just increase the

> amount people will gain from these projects just by giving them more of a free choice to do things

Student C found himself in a more strictly regulated project. The project was again developed with an authentic end product in mind (in this case a water management plan for a local water source), but the breakdown of activities was predetermined, and there was little differentiation in roles. While in some teams this led to a collaboration element that was valued, Student C and his team felt restricted, and quite early in the project began to rebel in subtle ways. Rather than investing in the project, they chose to cut corners with their work to meet the minimum standard, but not to make meaningful advances to their understanding or skills. Because of this, Student C and his team were far less likely to describe themselves as scientists, or to feel they were gaining skills relating to authentic scientific investigation.

A sense of autonomy does not exclusively derive from the design of the project, however; environments conducive to student autonomy have links to student perception of competency in the learning environment (Hatlevik, Throndsen, Loi, & Gudmundsdottir, 2018) and so it is perhaps unsurprising that there was ample evidence that when students felt less capable, they also felt less able to control the outcome of their projects. Student A for example had expressed relatively high Physics and Maths related self-efficacy before the start of his project and also more regularly reported feeling his choices were impacting the outcome of his project. I do not mean to imply that it is necessarily the self-efficacy that feeds into the autonomy, or that the reverse is true. Challenge project observations suggest the flow was multidirectional, with autonomy and self-efficacy sustaining each other with support from multiple sources.

As in the case of autonomy above, sources that tend to detract from self-efficacy tended to have negative effects on student perceptions of the project and of themselves. Student D, who started her Computer Science project. She had been informed early on in the project by her instructor that the majority of their work would be copy-paste, and that she would finish with enough time to pursue expansions to the project. A particular issue, relating to the way in which the data offered to them was converted for use with the software, meant that for several weeks, students were stuck trying to troubleshoot through a single issue. Under most circumstances, these students were accustomed to request assistance from their instructor, or perhaps one of their more adept peers. In this instance, none of these measures brought a solution, and the frustration was notable:

> Student D: it's kind of frustrating at the minute we can't do anything with it even [instructor] doesn't understand, the mentors obviously don't really understand so it's not as if anyone who's there can really help us

In this instance, the perceived lack of support was the biggest detractor from student attitudes towards the project. The students had until this point been able to make use of documentation, advice from their peers, help from industry mentors, and a healthy dose of trial and error to complete their tasks. Because of the specific nature of the task, such strategies were not successful in this case. As students became increasingly aware of their own limitations in this area, the majority of the class grew impatient and began to question their abilities to complete the project independently. When the developers offered a solution and work continued, there was a rapid return to optimism and language in general improved, as did the sense of autonomy and agency.

8.8 Applicability

Another key element of a successful PBL project is that it provides experience and knowledge that is applicable outside the project space. While the authenticity described earlier suggests that this should focus mainly on the transferrable, twenty-first century skills necessary in industry, this does not mean that the project need not connect to current course content as well. While many STEM-based PBL projects may take place as part of the instruction for an individual subject such as Physics, Chemistry, or Maths, the Challenge project exists as a separate learning opportunity for the students, many of whom may be enrolled in different ranges of courses. This makes designing projects that are truly interdisciplinary and applicable across STEM subjects a necessity. The Hydropower dam project for example required each team to gather evidence relating to stream conditions, topography, and soil content, calculate potential power output for various options, and design a final product base. Student A in particular found this project quite relevant to his interests and his studies:

> Student A: I would say it's um a bit more a bit better because it's more realistic and like for example our [previous] project we were building the perfect mouse house which I don't really see how it it's going to affect us to much it's going to if anything it's going to affect the company so their mice is fine but for example this [company's] if we were building a dam which any dam company would do and the fundamental basics are actually building a dam is stretching into other topics as well which I really like so
>
> R: What other topics?
>
> Student A: So other topics as in life so for example Physics or like uh when we were doing the angle of the sand so like how much do we press down until it crumbles down so that was a it was like it was about the dam project but like we learned many more other things from it so
>
> R: Okay kind of jumping on that Physics idea, if you'd been taught a lot of those concepts in a Physics class so you're just in a Physics classroom doing that um how does the way that you feel about this compare you think you'd feel?
>
> Student A: I think it would be tough to understand and definitely more boring because like it's we most of us are going to think 'well when am I going to use this when am I going to press it down blah blah blah' when we're building our dam project we wouldn't ask ourselves that because we know we're doing this to build a strong base for the dam but so it's it it I think it's better to use I think this Challenge was a great way to put lessons and other stuff together

Across all three projects observed at Riverside College, the students were unanimous that connections between the projects and their coursework or career aims was a critical factor in determining their enjoyment.

This does not mean that the project should be limited to the content contained within the course syllabi for any of the related subjects. While the ability to take knowledge from the Challenge back to the classroom was universally welcomed, Student B (from the Computer Science project) felt that one of the benefits of the Challenge project was that it was "greater than lessons":

> R: Yeah um if you were to be put in charge of designing these for next year let's say [Challenge coordinator] comes to you and says 'you're running this' um what would would you design them specifically to be separate from the lessons or would you want some of that overlap?
>
> Student B: I would want some of that overlap actually because you don't want to go into it not knowing anything you want to go into it knowing a little bit of something and then expand so I think I'd have like a little bit of overlap but not too similar to like anything you do in lessons cause otherwise it's kind of like repetition obviously kind of gets into your head but for exploring I feel like Challenge is for exploring new ideas and new themes

What overlap occurs course depend not only on the project, but on the subjects in which the student is enrolled. The experience of Student A, at GCSE (14–16), is different from those of Students B, C, and D enrolled in A Level or BTEC (16–18) courses. All of these will differ from students following other national curricula or the International Baccalaureate as well. The number of subjects, or the ways in which they are structured, may go a long way in determining how student experience impacts the project or *vice versa*. Student D for example felt that certain courses prepared students for the type of thinking necessary in successful programming, suggesting that "...if one of them does like Maths maybe and Physics it would be a lot easier maybe to kind of get that logical thinking cause they kind of already have it". This would seem to reveal a more established awareness of the importance of interdisciplinary thinking, possibly as the result of previous projects. Student A felt similar, believing a diverse STEM background from his coursework would make him better able to cope with the project, stating that "the energy of Chemistry and Physics is more like the world energy and how we can like produce it for our welfare,"which would be the focus of his hydropower project.

8.9 Exposure to Industry/Experts

An extension of the applicability feature discussed above, the ability of a project to act as a catalyst for the careers of its participants is also highly valued. When speaking with the Parkside College students, I was unsurprised to learn that each of the participants in the focus groups stated that they felt more invested in a project that involved useful industry contacts in an area of interest to them. One of the projects I observed there, held in conjunction with a local group of engineers, featured a competitive aspect with a judge panel of engineers from the company, including one judge who was an alumni of the school in question completing an apprenticeship with the company. His apprenticeship had been the result of his team winning the same competition a few years previously, and so the students were able to see for themselves the direct link from their project to their career paths.

Returning to Riverside College, students had fewer direct examples of their projects leading to careers, but the possibility was still an attractive one to many of the students. Student D for example was highly interested in pursuing a career in the industry surrounding her project, and thus the collaboration was for her not just about completing the project, but about networking and displaying her skills as well. She maintained determination throughout the project to implement a particular feature in her project because it "…was their new feature so I really wanted to try and get that in be able to use it kind of show it off cause I knew that would impress them…". Other groups abandoned certain goals because of their difficulty level, but Student D's strategy was to successfully implement something "they were still trying to get the hang of it and because if we managed to do it in eight weeks not even that like four umm that would be really impressive". Her strategy in the end paid off, with her team receiving multiple compliments from the expert judges' panel, and as she also pointed out, an award for her CV that showed she was capable of high achievement in the field.

Even students who did not plan on careers related to their specific projects found the opportunity to work with industry experts an appealing one. Student C for example found that his ability to consider evidence was tested by his water management project:

> Student C: He [Conservationist] was giving this great passionate good speech about why we should save [local waterway] and I got to the end and I was like 'I'm totally with this guy has completely got my support, I'm going to go home and research this' and then the [Water Company Representative] got up and she started talking she's like 'it's not that I don't care about this but I also have these other priorities on my plate' at which point I was able to be like 'oh. Okay right' and then I got a bigger understanding…personally on a wider scale

Student C was faced with two experts with different focuses, and given the task of determining what his priority would be, and to come up with a compromise if possible. This required listening, taking careful notes, some verification and research, and then argumentation skills as he debated with his peers in a panel discussion later in the day. Though his project featured fewer sessions with external experts, each time he interacted with one he expressed an appreciation for the new perspectives they offered him, and over time began to communicate more as a scientist than as a student of science.

8.10 Implications for Teaching

Just as each project is different, so too is every learner engaged in the completion of them. This makes the task of determining an effective strategy for developing a successful project a challenge. There are certain elements that each of the projects had that resonated well with the students, and elements that tended to detract from their experience. I offer these here as a list of elements to consider in the design of an authentic PBL project.

1. Authenticity. This may seem the obvious point, but consider again for a moment the hypothetical dam project. A project that is truly authentic will often allow for greater student autonomy, the differentiation of roles, and memorable experiences that students can use to connect the classroom to the real world. Students are also more likely to make links from subject to application if an effort is made to allow for the application of classroom-based skills and understandings. Authenticity in STEM in particular will mean considering the skill sets expected once the learners enter industry, and designing a project that allows them to develop these. This was particularly evident in the Hydropower dam and Computer Science projects at Riverside College, but elements of authenticity could be seen in all projects observed.
2. Flexibility. Students across all observed projects felt that having goals that could be adapted to suit their interests and abilities was a key to their enjoyment and success. For some, this meant an open-ended problem to be solved in a number of different ways. In Riverside's hydropower project, it meant allowing students to set the specifications and then deal with the hypothetical real-world consequences as they saw fit. It was only in projects where students perceived that their input was being disregarded that significant disinterest was seen. As was discussed in the introduction, PBL is most often concerned with the creation of a tangible artefact, so allowing students some ownership of that artefact may greatly enhance their enjoyment of and engagement with the project.
3. Autonomy. Beyond just a sense that they were in control of the final product, the students universally expressed desires to be in control of the tools they would need to meet their goals. For Student C, this was as simple as being given a syllabus in advance so he could plan for his final product. For Students B and D, in the Computer Science project, this was information on how to manage and process the data. Student A felt that he had been given not only the tools to succeed, but the flexibility to use them as he wished, and expressed the greatest confidence in his ability to self-regulate throughout the project. It was when students felt empowered by the resources available to them, they expressed a greater sense of autonomy, and usually also an increased appreciation for the project as a whole. Autonomy also means students have more of an opportunity to develop skills relating to planning, problem-solving, teamwork, and communication.
4. Relevance. While it was noted in the introduction that projects need be central rather than peripheral to the curriculum, the degree to which this is so matters to the success of the project. In Riverside College, where the Challenge projects are designed and marketed as a way for students to develop transferrable, STEM skills that can be brought back into any classroom subject, students were receptive even to projects outside their defined areas of study. Where a PBL project has more direct ties to a single subject, the priority is to balance the inherent value of the project as a way to reinforce curricular concepts with the ability of the project to broaden the skill set of the learner.

5. Interdisciplinarity. Underpinning all of the previous elements, and the projects themselves, is the interdisciplinarity that allows for students engaged in them to collaborate and apply their knowledge from their diverse backgrounds to develop creative solutions to real-world problems. Interdisciplinarity can also in itself lead to authentic learning contexts (Ledoux & McHenry, 2004) and fits quite comfortably within the constructivist frameworks that support project-based learning as a teaching and learning method. Interdisciplinarity in these projects does not have to be forced; in several of the observed projects, it was the students who chose to incorporate an interdisciplinary approach to solve problems and complete their tasks. As mentioned previously, a STEM education is more than the sum of its parts, and so is this true of the interdisciplinary nature of the STEM project. The nature of a well-designed STEM project task will require students to draw on skills and knowledge that span the STEM disciplines and interlink them in order to succeed.

It is worth noting that the schools discussed in this chapter are exceptional cases themselves; from their inception they were intended to facilitate interdisciplinary, STEM-focused projects with collaboration from local industry. While UTC's in England, and initiatives such as the Harmony Public Schools (Sahin, 2015) in America, may continue to be the exceptional standard, this does not mean that existing projects cannot be modified to focus on learning for, rather than learning from, the real world. Lessons from these models can be used to adapt projects within single subject classrooms. Returning to the hypothetical dam project, a prompt such as 'consider the social, economic, and environmental implications of building a large-scale hydropower dam' would require students to access information from multiple disciplines. An instructor need not have specialist knowledge in every area to be able to offer meaningful guidance either; where industry collaboration is not possible, peer collaboration almost certainly is. I recall an incident from my own classroom days where my teaching team was running a project on animal migration, only to discover through chance that at the same time the social sciences department were focusing on complementary issues.

8.11 Final Thoughts

This is not meant to be an exhaustive how-to guide for project-based learning nor is it meant to define it, or even to clarify what it means for an assignment to be PBL. It is instead meant to convey the student experience in a variety of PBL projects, beyond their measurable learning outcomes. The students represented here, like the schools, may be considered outlier cases due to their deliberate enrolment in a programme that includes these PBL components, but this does not diminish the value of their stories. Each student came into these UTC's with academic and career aspirations, and thus their perspectives still represent a diverse range of students and interests. These perspectives were not the aim of the interviews; the primary focus had been on learning outcomes. The degree of agreement regarding experiences across unconnected projects, between students who had no interactions and little in common, made it impossible to ignore this angle, and thus this chapter was born out of a desire to relate the stories of the students as people as well as learners. What all these stories had in common was that when students were given the chance to be creative, to think collaboratively and interdisciplinarily, and to assert their autonomy over their outcomes, they were both motivated and engaged. PBL projects will take many forms, some more authentic than others, but by considering the elements described in this chapter, it is possible to ensure that students are given the opportunity to work in ways that connect their learning to their lives in meaningful and productive ways.

References

Bender, W. (2012). *Project-based learning: Differentiating instruction for the 21st century.* United States: Corwin

Breiner, J. M., Harkness, S. S., Johnson, C. C., & Koehler, C. M. (2012). What is STEM? A discussion about conceptions of STEM in education and partnerships. *School Science and Mathematics, 112*(1), 3–11.

Brown, R., Brown, J., Reardon, K., & Merrill, C. (2011). Understanding STEM: Current perceptions. *Technology and Engineering Teacher, 70*(6), 5.

Canelas, D. A., Hill, J. L., & Novicki, A. (2017). Cooperative learning in organic chemistry increases student assessment of learning gains in key transferable skills. *Chemistry Education Research and Practice, 18*(3), 441–456.

Carvalho, A. (2016). The impact of PBL on transferable skills development in management education. *Innovations in Education and Teaching International, 53*(1), 35–47.

Dobrin, J. (2020). *Investigating learning in secondary science students engaged in project-based learning*. [PhD thesis]. University of Cambridge

Fink, L. D. (2013). *Creating significant learning experiences: An integrated approach to designing college courses*. San Francisco, CA: Jossey-Bass.

Hasni, A., Bousadra, F., Belletête, V., Benabdallah, A., Nicole, M.-C., & Dumais, N. (2016). Trends in research on project-based science and technology teaching and learning at K–12 levels: A systematic review. *Studies in Science Education, 52*(2), 199–231. https://doi.org/10.1080/03057267.2016.1226573

Hatlevik, O. E., Throndsen, I., Loi, M., & Gudmundsdottir, G. B. (2018). Students' ICT self-efficacy and computer and information literacy: Determinants and relationships. *Computers & Education, 118*, 107–119.

Herrington, J., & Oliver, R. (2000). An instructional design framework for authentic learning environments. *Educational Technology Research and Development, 48*(3), 23–48.

Kokotsaki, D., Menzies, V., & Wiggins, A. (2016). Project-based learning: A review of the literature. *Improving Schools, 19*(3), 267–277. https://doi.org/10.1177/1365480216659733

Ledoux, M., & McHenry, N. (2004). A constructivist approach in the interdisciplinary instruction of science and language arts methods. *Teaching Education, 15*(4), 385–399.

McCrone, T., White, R., Kettlewell, K., Sims, S., & Rush, C. (2019). *Evaluation of university technical colleges*. Slough, UK: NFER.

Mills, J. E., & Treagust, D. F. (2003). Engineering education—Is problem-based or project-based learning the answer. *Australasian Journal of Engineering Education, 3*(2), 2–16. http://www.aaee.com.au/journal/2003/mills_treagust03.pdf

Prince, M. J., & Felder, R. M. (2006). Inductive teaching and learning methods: Definitions, comparisons, and research bases. *Journal of Engineering Education, 95*(2), 123–138. https://doi.org/10.1002/j.2168-9830.2006.tb00884.x

Sahin, A. (2015). STEM students on the stage (SOS): Promoting student voice and choice in STEM education through an interdisciplinary, standards-focused project based learning approach. *Journal of STEM Education, 16*, 3.

Thomas, J. W. (2000). *A review of research on project-based learning*. San Rafael, CA: Autodesk. http://www.k12reform.org/foundation/pbl/research

University Technical Colleges (UTC). (2019). *The UTC story*. Retrieved from https://www.utcolleges.org/the-utc-story/

Jessica Dobrin is a teacher of Science and Chemistry, and a researcher in oject-based learning PBL. She completed a PhD in Education at the University of Cambridge, where her research focused on schools using PBL to facilitate real-world learning in STEM. She currently teaches at Dulwich College, Beijing.

Chapter 9
Collaboratively Learning to Teach STEM: A Model for Learning to Integrate STEM Education in Preservice Teacher Education

Sevil Akaygun and Fatma Aslan-Tutak

9.1 Introduction

As world is becoming more globalized, it is becoming more interactive and collaborative, as well. This transformation is also seen in STEM (Science, Technology, Engineering, Mathematics) education, whose importance is increasing in that individuals with STEM skills are better able to adapt to the every-changing world. Education systems around the world have therefore started paying attention to and including STEM education into their programs. In this context, preparing teachers who will implement STEM education is as crucial as the STEM education itself. They need to be experts in their specific fields who can collaborate with educators in other STEM fields. This chapter describes the designed-based development of a module that introduces integrative STEM education that emphasizes meaningful interdependence among all the disciplines of STEM, to preservice science and mathematics teachers: Collaboratively learning to teach STEM (CLT-STEM).

9.1.1 STEM Education

STEM education initiatives started in the twentieth century in the United States, from where it spread to other countries (National Research Council, 2010). Some of the early research on STEM education focused on the increasing need on the job market for highly skilled people in STEM areas, but the focus has expanded to include STEM in education systems as early as prekindergarten years. Attempts to implement STEM education at preschool and primary levels opened up a

S. Akaygun (✉) · F. Aslan-Tutak
Bogazici University, Istanbul, Turkey
e-mail: sevil.akaygun@boun.edu.tr

V. L. Akerson, G. A. Buck (eds.), *Critical Questions in STEM Education*, Contemporary Trends and Issues in Science Education 51,
https://doi.org/10.1007/978-3-030-57646-2_9

discussion on integrating four subject areas: science, technology, engineering and mathematics in a meaningfully interdependent way. The fundamental underlying rationale for integrating them has been that the whole is, in effect, more than the sum of the parts. In that matter, the individual areas of STEM are related to each other and integrating them in educational settings calls for interdisciplinary work of areas.

Thus, the discussion on STEM education has been on how STEM areas can be integrated for STEM education in the twenty-first century. Honey, Pearson, and Schweingruber (2014) indicate that "working in the context of complex phenomena or situations on tasks that require students to use knowledge and skills from multiple disciplines" (p. 52). One STEM education initiative in the U.S. was the STEM Taskforce Report (2014), which recommends the integration of areas by focusing on problem-based learning to deal with real-world problems, where subject areas are not isolated. An immediate change was witnessed in science curricula, with the inclusion of engineering processes. Again in the U.S., Engineering, Technology and Applications of Science Standards, which were connected to core disciplines and ideas and crosscutting concepts were added to the Next Generation Standards for Science (NGSS Lead States, 2013). A similar initiative has been launched in Turkey, which has been added to the Turkish National Science Curriculum for grades 4–8 (Ministry of National Education (MoNE), 2018) under the heading of science, engineering and entrepreneurship applications which is also reflected in CLT-STEM module developed by the authors.

To investigate STEM education practices of different countries, Ritz and Fan (2018) surveyed technology education scholars from 20 different countries: Australia, Belgium, Canada, China, England, Finland, France, India, Ireland, Israel, Japan, the Netherlands, New Zealand, Scotland, Singapore, South Africa, South Korea, Sweden, Taiwan, and the United States. The results showed that the Netherlands and South Korea had been working on solely integration of STEM, but England, Ireland and Japan had been focusing on including STEM only as an enhancement. The remaining countries were categorized as "exploring options of integrative STEM and the improvement of the stand-alone individual STEM subjects" (p. 445).

The term "STEM education" generally refers to the integration of science, technology, engineering and mathematics, but it sometimes simply alludes to the provision of limited support for each of these areas individually. Thibaut et al. (2018) identified three challenges for full integration of the STEM areas: finding resources, enhancing teachers' knowledge, and arriving at a consensus on the degree of integration. After a systematic analysis of published research on STEM education, they identified five key principles for integrating STEM into secondary schools: integration of STEM content, problem-centered learning, inquiry-based learning, design-based learning and cooperative learning. Integrating STEM education is far from a trivial task in itself; preparing knowledgeable with the right skills to implement STEM education initiatives is an additional challenge.

9.1.2 Integrating STEM Education into Teacher Education

The desired effect of integrated STEM education in K-12 can only be achieved by competent science and mathematics teachers who can successfully integrate STEM into their regular classes. This can be accomplished either by providing professional development for inservice teachers or by introducing STEM education into preservice teacher education programs. In both cases, the training should involve experiencing, practicing and guiding, with a focus on both implementation and theoretical underpinnings. Such training enhances the qualifications of STEM teachers.

Numerous countries have recognized the need for qualified STEM teachers. In the United States, for instance, the National Science Board (NSB) highlights what teachers need in order to effectively teach STEM and draws attention to the importance of attracting, preparing, and retaining qualified and committed teaching candidates (NSB, 2007, cited in Stohlmann, Moore, & Roehrig, 2012). European Schoolnet, the community for science education in Europe, reports the results of a study conducted in 14 European countries, including Turkey, that highlights the need to attract more teachers to STEM education, from primary school to adult education (European Schoolnet, 2018). With this in mind, to improve teachers' competencies in STEM education, educational systems need to provide special training, support the publishing of pedagogical content, and facilitate the sharing of best practices for enhancing STEM education by focusing on teachers' needs through a common framework and each country's own strengths.

All around the world, various inservice and preservice teacher education programs have been developed. Some programs for inservice teachers have been held during the summer breaks (Nadelson, Seifert, Moll, & Coats, 2012) and focused on knowledge, instruction, and efficacy. Such programs increased teacher confidence, improved content knowledge, and made better use of community resources in the integration of STEM into classroom practices (Baxter, Ruzicka, Beghetto, & Livelybrooks, 2014; Nadelson & Seifert, 2013). Professional development programs were conducted over a longer term. A one-year integrated engineering program was found to be effective in implementing engineering practices into classroom contexts (Guzey, Tank, Wang, Roehrig, & Moore, 2014). Similar STEM education programs developed for preservice teachers focused on content and inquiry, and these were also found to be effective for improving teachers' competence and confidence in STEM (Murphy & Mancini-Samuelson, 2012), their perception of the feasibility of content integration (Berlin & White, 2012), and their understanding of the nature of STEM (Aslan-Tutak, Akaygun, & Tezsezen, 2017; Yıldırım & Türk, 2018).

The need to improve the quality of STEM teacher education inspired researchers to develop and investigate the effects of implementing different methods of integrating STEM education into preservice teacher education. These include design-based science education (Bozkurt Altan, Yamak, & Buluş Kırıkkaya, 2016), laboratory-integrated STEM applications (Gökbayrak & Karışan, 2017; Yıldırım & Altun, 2015), and technology-integrated STEM applications (Alan, Zengin, & Keçeci, 2019). Bozkurt Altan et al. (2016) implemented design-based science activities to preservice science teachers during a one-semester laboratory course where they worked on four different design projects. In semi-structured interviews, preservice teachers described their experiences as permanent learning and highlighted the value of experiential learning; they also indicated a need for meaningful learning, improved science process skills, and support for higher-order thinking. They also mentioned increased motivation, positive attitudes and responsibility for learning. In another study, Gökbayrak and Karışan (2017) report that implementing STEM-based activities for 13 weeks improved the science process skills of preservice teachers. Using Algodoo software, Alan et al. (2019) implemented technology-integrated STEM applications to support the preservice science teachers in the development and use of problem-solving activities. In their study, 31 preservice science teachers were first introduced to STEM education, integrated pedagogical knowledge, and the importance of integrating different disciplines. Over the next five weeks, they designed simulations and finally used their simulations in microteaching sessions. It was reported that the problem-solving skills of the preservice teachers improved significantly more than their counterparts in the control group.

With respect to the design of professional development programs in STEM education, several researchers have suggested key characteristics of an effective teacher training program in STEM education. Stohlmann, Moore, and Roehrig (2012), for example, cited support, teaching, efficacy, and materials. They emphasize the importance of supporting teachers in STEM integration, given that it is a fairly new area for them, and letting them practice their area of expertise while integrating STEM. They also recommend using various teaching methods to enhance teachers' self-efficacy beliefs, and using appropriate materials in design activities so that students can better understand STEM.

This chapter describes the development and evaluation of a research-based STEM education module "Collaboratively Learning to Teach STEM" (CLT-STEM) with the approach of STEM as having meaningful interdependence among the disiplines. CLT-STEM was developed over three-phase iterative design-based instruction for the teaching methods courses for preservice teachers in two STEM disciplines, chemistry and mathematics. The preservice teachers worked collaboratively in small groups on three different STEM activities that integrated chemistry, mathematics, technology and engineering. Having reflected on how STEM was integrated, they developed a STEM project suitable for implementing in their future classes. To this end, we aimed to identify the key components of learning integrative STEM education for the interntional audience of STEM education practioners and researchers.

9.2 Methodology

The theoretical framework adopted in this study was social constructivism, which accepts individuals as unique human beings who make meaning from their own experiences as they interact and collaborate with others in a social environment (Vygotsky, 1978). In our case, the learners (preservice teachers) actively learned the notion of integrative STEM education while collaboratively working on various tasks that involved different dimensions of STEM education. After completing each task, they reflected on their own practices, allowing them to connect the new information with their prior knowledge and hence construct their own understanding. In other words, they learned what integrative STEM education was through the lens of a social constructivist approach.

We used a design-based approach, which is a "systematic methodology [that] aims to improve educational practices through iterative analysis, design, development, and implementation, based on collaboration among researchers and practitioners in real-world settings (Wang & Hannafin, 2005, p. 6)." We developed an integrative STEM education module that we called "Collaboratively Learning to Teach STEM" (CLT-STEM) for preservice chemistry and mathematics teachers. The module was implemented and evaluated in three cycles, with the second and third implementations being iterations of the one before it, following the designed-based methodology.

9.2.1 *Participants*

The study was conducted at a public university in Istanbul by the university faculty members who taught the teaching methods courses for preservice teachers of mathematics and chemistry. Similar to many other secondary school teacher education programs, prior to taking the methods course at their final year in the program, the participating preservice teachers had taken, on average, 15 area specialization courses (e.g. chemistry courses in the case of chemistry preservice teachers) and one prerequiste methods course that focused specifically on teaching science and mathematics. Thus, findings of this study may be meaningful for secondary teacher educators from other countries. A total of 138 preservice teachers participated in the study in the 3 years that it was implemented: 55 were in chemistry and 83 were in mathematics. The breakdown was as follows: in 2014, there were 50 (18 chemistry and 32 mathematics); in 2015, there were 48 (22 chemistry and 26 mathematics); and in 2017, we had 40 (15 chemistry and 25 mathematics) preservice teachers. Participants were asked to form groups of 3–4 with only one stipulation—that they should have at least one chemistry preservice teacher and at least one mathematics preservice teacher in each group. The aim was to have participants work with peers from another discipline while giving them the opportunity to form their own groups.

9.2.2 *Procedure*

The CLT-STEM module, developed and implemented in three academic years (2014, 2015 and 2017) was designed to have chemistry and mathematics preservice teachers experience integrative STEM education activities suitable for

Table 9.1 The CLT-STEM Module Schedule

Time	2014	2015	2017
Week 1	Administration of PreTest	Administration of PreTest	Administration of PreTest
	Introduction to STEM Education & Discussion	Introduction to STEM Education & Discussion	Introduction to STEM Education & Discussion
Week 2	Edible Cars Activity in class application	Announcing Ocean Color Activity & out-of-class group work	Announcing Ocean Color Activity & out-of-class group work
Week 3	Ocean Color Activity in class application	Ocean Color Activity in class discussion	Ocean Color Activity in class discussion
Week 4	Nanotechnology and Origami Activity in class application	Edible Cars Activity in class application	Edible Cars Activity in class application
Week 5	Out-of-class group work on final project	Nanotechnology and Origami Activity in class application	Sailing Boat Activity in class application
Week 6	Integrative STEM Education Exhibition	Integrative STEM Education Exhibition	Integrative STEM Education Exhibition
	Administration of PostTest	Administration of PostTest	Administration of PostTest

secondary school students and then develop their own integrative STEM education activity. The researchers adopted selected science, mathematics and STEM education activities to allow participants to experience the integration of STEM areas (Aslan-Tutak, Akaygun & Tezsezen, 2017). The module was implemented as part of a regular teaching methods course for preservice science and mathematics teachers. The duration of the module was six weeks; each week there were two lecture hours (100 minutes). Table 9.1 summarizes the content of the module for each of the 3 years.

As can be seen in Table 9.1, there were some changes over the years about implementations of CLT-STEM. After evaluating the first implementation, the order of the first and second activities was reversed to allow more interaction and easier collaboration between the members of the respective groups. After the second implementation, the final activity was replaced with an activity developed by the preservice teachers themselves (Sailing Boat Activity) so as to let third year participants to experience an activity developed by their peers. They then reflected on the nature of the activity and its potential for in-class use.

In each implementation cycle, the participants read two introductory articles about STEM education: Dugger (2010) and Laboy-Rush (2011). This introduction was followed by three activities to be completed in groups. During the next four weeks, the preservice teachers worked collaboratively in their self-selected groups on STEM activities; these were related to Edible Cars, Ocean Color, Nanotechnology and Origami. Each activity integrated chemistry and mathematics meaningfully and included technology (as a tool or process) and engineering (in terms of design principles or process). At the end of each activity, they discussed how STEM had been integrated and how they might implement such projects in their future classes.

Fig. 9.1 Edible cars constructed by preservice teachers

Individually, they wrote their reflections on the implementation of thees activities, focusing on their own experiences and on the collaboration.

The first activity, Edible Cars, was adapted from an activity called edible car contest, which had been implemented during the National Engineering Week (Perez, Gibson, Opsal & Lynch, 2011). During this activity, the preservice teachers worked in groups, first to design their edible car on paper and then constructed it from items such as cookies, fruits, crackers, etc. The car had to run on a one-meter ramp without breaking down. Then, the groups calculated the average speed of their car and discussed the problems they experienced and developed solutions. Figure 9.1 shows some of the edible cars built during the activity (Tosmur-Bayazıt, Akaygün, Demir & Aslan-Tutak, 2018). During the Edible Cars activity, the preservice teachers needed to integrate their knowledge of chemistry and mathematics. For instance, when they used a cucumber for the body of the car and cracker sticks for the axles; the crackers absorbed water from the cucumber, got softer and broke down; they needed to find a solution for this problem. They usually covered the crackers with another food item such as jelly candies to create insulation to prevent absorption of water. In that case, they needed to know that jelly candies do not dissolve in water, and they needed to calculate the minimum amount of jelly candy to cover the whole or some parts of the surface of the crackers.

The second activity, Ocean Color, was adapted from Project Oceanography (Project Oceanography, 1999), which introduces the fundamental scientific concepts of oceanography, including photosynthesis, concentration, and light. During the activity, the preservice teachers again worked in groups to develop a model that would predict the factors affecting the color of oceans. The instructions for the activity were given in class, and the preservice teachers had 1 week to work in groups outside class and do research using the online sources provided on the activity sheet (Aslan-Tutak, Akaygun, & Tezsezen, 2017). They had to use their chemistry knowledge in terms of considering the solubility of CO_2 gas in water and

Fig. 9.2 Structural model constructed in the Nanotechnology and Origami Activity

their knowledge of mathematics to include the correct amount in their model as a variable.

The third activity, Nanotechnology and Origami, that was adapted from the Science Buddies website (Science Buddies Staff, 2015), aimed to model the structure of nanomaterials through paper folding. Each group was asked to construct a bridge modelling the allotropes of carbon using a maximum of 10 sheets of paper that measured 25 × 25 cm. They worked in groups to design a structural model that could support the maximum weight as they hang a paper cup bound to a ribbon on their model, as illustrated in Fig. 9.2 (Aslan-Tutak, Akaygun, & Tezsezen, 2017). In this activity, the preservice teachers needed to activate their chemistry knowledge in terms of carbon allotropes and their mathematical knowledge in order to fold the sheets of paper in a way that represented these allotropes as a geometrical shape.

Upon completion of all three activities, participants were asked to develop or adapt an integrative STEM education activity suitable for their future classes. The groups then presented their projects to their peers in a STEM exhibition and received feedback from both their teacher educators and their peers. Prior to and after the implementation, the preservice teachers answered questionnaire with open-ended items designed to elicit their views of integrative STEM education. Semi-structured interviews were then conducted with selected participants. The group projects were evaluated and individual reflections were examined, all analyses combined contributed to the evaluation that year's implementation and informed revisions for future iterations.

The preservice teachers' responses to items on the questionnaire, administered both before and after each year's implementation of the CLT-STEM module, revealed their level of awareness about STEM education and the interviews provided an opportunity for them to suggest revisions to the module. The iterative technique of revising and testing the CLT-STEM module was in tune with the nature of developing design-based instruction. The module was implemented three times in 3 years: 2014, 2015 and 2017. The focus on the integrative approach to STEM education, including all individual disciplines of STEM in a meaningfully interdependent way, was maintained in all implementations, yet several modifications were made to facilitate the preservice teachers' collaboration. For example, in the first implementation of the module, all three activities were executed in class, and the integrative STEM projects were developed outside class. The interviews revealed that the preservice teachers had difficulty in collaborating outside the class, so in the second year, the Ocean Color Activity was changed to a "do-it-outside, discuss-it-in-class" activity: they first worked in groups outside the class, and then discussed their work with their classmates in class. Written reflections and interviews from the second year showed that the collaboration problem had been resolved; as they were collaborating with their peers outside class from the very first week of the implementation. They still needed encouragement to implement STEM education in their future classes, however. For this reason, for the third implementation, one of the activities, the Nanotechnology and Origami Activity, was replaced with the Sailing Boat Activity that had been developed by one of the groups in the first year's implementation.

9.2.3 Data Sources and Analysis

Data were drawn from various sources in order to address different research questions. In this chapter, the authors focus on participants' answers to two open-ended questions on definition of STEM education and the methods used during STEM education in order to provide information on design-based nature of the module for the international audience of STEM education practioners and researchers. The open-ended questions that addressed these issues were posed in a questionnaire that was administered twice—once at the beginning of the implementation and again at the end, and findings from the pre and post individual interviews. Participants' answers to the questionnaires were first openly coded for content analysis. The categories emerged for the definition of STEM education were (i) instruction to capture students' interests, (ii) comprehensive instruction, (iii) teaching subject areas together, and (iv) teaching subject areas in an integrated way. Answers which had statements that fell into more than one category were assigned multiple codes, so the number of categories that emerged from data is greater than the number of participants. One participant's answer, for example, was "*A program to be implemented with the purpose of explaining how mathematics is used in science, technology and engineering.*" This was coded as "comprehensive instruction", whereas "*Using these four areas in a relation while teaching a topic*" was coded as "teaching subject areas together". The differentiation between the category of "teaching subject areas together" and the category of "teaching subject areas integrated" can be explained by a sample of a participant's answer coded as "teaching subject areas integrated", "*using more effective methods by using science, technology, engineering and mathematics integrated.*"

The researchers also conducted interviews with some preservice teachers from each subject area (chemistry and mathematics). In those interviews, one of the questions was about the participants' experiences of collaborative learning for integrated STEM education. Again, the answers were first open-coded, and then categories emerged. The responses to questions about the collaborative nature of interdisciplinary work fell into three categories (i) knowledge, (ii) group dynamics and (iii) non-academic factors.

9.3 Findings

9.3.1 *Participants' Definition of STEM Education*

The responses of participants to the questions that explored their definition of STEM education fell into four categories, as mentioned above. Based on the descriptive analysis, a frequency table for these categories were prepared. A comparison of participants' responses showed that participants' definition of STEM education evolved from "capturing students' interest or teaching areas together" and "comprehensive instruction" to "integration of the subject areas". The number of occurrences of integration of areas from participants' answers was found to be doubled by increasing to 24 in the PostTest from 12 in the PreTest as seen in Table 9.2.

Furthermore, considering these categories from integrative STEM education perspective yielded to grouping them in an ordinal perspective, from "no integration but simple interesting activities" to "teaching subjects together", and then "integration of subject areas". This ordinal coding schema allowed researchers to use the Wilcoxon Signed Rank Test, which was used for the analysis of frequencies. The results were found to be statistically significant ($Z = -2.748$, $p = 0.006$).

Table 9.2 The Frequencies of Codes found in the Categories for the Definition of STEM Education

STEM Education Definition Categories	Frequency of codes in the PreTest	Frequency of codes in the PostTest
Instruction to capture students' interest	4	2
Comprehensive instruction	9	3
Teaching subject areas together	20	20
Teaching subject areas in an integrated way	12	24
Total number of codes	50	51

Table 9.3 Frequency of the Codes found in the Categories for Methods Used in STEM Education

Methods used in STEM Education	Frequency of codes in the PreTest	Frequency of codes in the PostTest
Inquiry	31	47
Modelling	5	14
Engineering processes	7	17
Total number of codes	43	78

9.3.2 *Participants' Views on Methods for STEM Education*

In the questionnaire, preservice teachers were also asked about methods used in STEM education. Their responses fell into three categories: inquiry, modelling, and engineering processes (including technology). Table 9.3 shows the frequencies of responses given in the PreTest and PostTest. The *inquiry* category included the codes of experimentation, investigation, and implementation; the *modelling* category included the codes of constructing models and projects; and finally, the *engineering processes* category included the codes of engineering/design processes and using technology.

Table 9.3 also indicates that the views of preservice teachers about methods used in STEM education changed towards inquiry-based, modelling-based and engineering processes-based STEM education from PreTest to PostTest. The change in the number of codes in each category, as well as the total number of codes, were found to be significant when compared with Wilcoxon Signed Rank Test ($Z = -4.455$, $p = 0.000$).

The purpose of the CLT-STEM module was to change participating preservice teachers' views of STEM education toward integration of areas by facilitating meaningful collaboration between subject areas (chemistry and mathematics, in this case). Thus, the researchers also investigated participants' views of collaborative work while experiencing CLT-STEM module.

9.3.3 *Participants' Views on Collaborative Work in STEM Education*

Participants' views on the collaborative nature of interdisciplinary work fell into three categories: (i) *knowledge*, (ii) *group dynamics* and (iii) *non-academic factors*.

(i) knowledge: The knowledge category consists of two sub-categories: *subject matter knowledge*, and *knowledge of the other discipline*. Participants discussed their knowledge in their own subject area, that is, mathematics preservice teachers' knowledge of mathematics and chemistry preservice teachers' knowledge of chemistry. They also talked about the knowledge of group members in the other discipline (e.g. preservice mathematics teachers' knowledge of chemistry). The analysis revealed, not surprisingly, that a participant's lack of subject matter knowledge diminishes the interdisciplinary work of the group, and that knowledge of the other discipline improved their work. Since the interdisciplinary activities in the CLT-STEM module required chemistry knowledge as well as mathematics knowledge, the group members had to explain chemistry and mathematics ideas to each other. Group members therefore had to depend on each other in terms of their expertise in the subject matter. Indeed this chracteristics of CLT-STEM reflects this research's approach to STEM education as meaningful interdependence among all disciplines of STEM.

Two mathematics preservice teachers highlighted *learning chemistry* while working with a chemistry group member, but another participant criticized her chemistry group member. For example, Osman reported having a *fruitful* group work and learning chemistry from his group member:

> While working with Berrin [chemistry group member] for our activity, I learned something about chemistry etc.

Similarly, Ilker's evaluation was positive:

> Actually she [the chemistry group member] knows what I don't know and I know what she doesn't know, and putting these together was a nice combination, if you ask me.

Kemal, on the other hand, criticized chemistry member of the group for her lack of chemistry knowledge:

> The whole-group classroom discussions were nice. I mean, like I said before, the feedback we gave to each other was good. In our group, our work with the group member [chemistry group member] was OK. But we did not get much help for the chemistry part. We had such a problem.

Kemal explained that because of the chemistry group members' lack of subject matter knowledge, they were unable to proceed in the CLT-STEM activities. Indeed, at the end of the CLT-STEM module, the STEM project developed by his group exhibited some misconceptions about chemistry. Similarly, some chemistry participants also reported that mathematics group members were unable to contribute to the integrative STEM education activity development.

(ii) group dynamics: Participants mentioned instances of not being able to work properly due to problems with the group not working smoothly in ways that would support collaboration. Those experiences were coded as *group dynamics*. One participant talked about a mathematics group member's lack of motivation and failing to attend group meetings outside the class. For example, Poyraz explained what had gone wrong with his group as follows:

> In general, this is not related to being a chemistry person or a math person. We had some sort of personal disunity.

Other participants, however, reported positive experiences. One pointed out, for example, that knowing other group members before the CLT-STEM module was helpful in building a collaborative group.

(iii) non-academic factors: The last category, *non-academic factors*, consists of such elements such as the groups' making arrangements for a time and place to meet and work. Participants who formed constructive groups reported having successfully resolved time and place issues. On the other hand, participants whose group had not been able establish a positive learning environment cited other non-academic factors. Poyraz, for example, discussed his experience of not being able to work well in a collaborative group. Some of his remarks were coded as group dynamics. His group had problems in arranging meeting times and communication, which resulted in their group not working together.

> We had difficulty in turning ideas in real world, especially because of communication and arranging meeting times.

In the interviews, participants who reported constructive interdisciplinary work experience talked about learning from each other and gaining a new perspective on investigating a phenomenon. On the other hand, two of eight participants expressed difficulty completing their interdisciplinary task in a group. All in all, participants' views on the collaborative aspect of integrated STEM education reflected possibilities and challenges of interdisciplinary work which may inform practioners and researchers of STEM education. As groups worked together on integrative STEM education activities and on developing an integrative STEM education activity, they depended on each other's discipline-specific knowledge to solve group dynamics issues as well as non-academic issues related to group work. These findings informed decisions about revisions to the CLT-STEM module for the next implementation in a way that would allow participants to have a more efficient interdisciplinary group work. All in all, the findings revealed the key features of the CLT-STEM module: integration, collaboration, skills needed for implementation, and learning to teach STEM.

9.4 Discussion and Conclusion

Given an increasingly pressing need for STEM education, many countries have been taking action to integrate STEM education into their science programs, and the need for incorporating STEM into teacher education becomes more evident. To this end, STEM education training programs have been developed for both inservice teachers (Berlin & White, 2012; Nadelson & Seifert, 2013) and preservice teachers (Alan, Zengin & Keçeci, 2019; Aslan-Tutak, Akaygun, & Tezsezen, 2017, Yıldırım & Türk; 2018). In those programs, teachers (preservice or inservice) are introduced to STEM Eductaion framework, principles and applications through collaborative work. To enhance STEM teacher education, various implementations of STEM education for preservice teachers have resulted in improvement in problem-solving skills (Alan et al., 2019), academic achievement (Yıldırım & Altun, 2015), science process skills (Gökbayrak & Karışan, 2017), learning, motivation, attitude, responsibility for learning - as self-reported (Bozkurt Altan et al., 2016). The growing body of knowledge in various countries and cultures contributes to the developments in STEM education research and practice.

With the purpose of contributing international discussions on STEM education, in this chapter, we have presented an integrative STEM education module, CLT-STEM, for preservice chemistry and mathematics teachers that was developed in three cycles, with the second and third being iterations of the previous cycle. Using design-based approach to develop CLT-STEM, the results obtained from one year's iteration was used to improve the implementation in the following year. Based on the analysis of preservice teachers' responses to questionnaires and their answers in interviews, the components that need to be included in the module have been recognized. Specifically, the chemistry and mathematics preservice teachers emphasized learning STEM education which involves integration of STEM fields while working on the tasks collaboratively. These STEM education-related tasks are typically designed to include the methods of inquiry, modelling and engineering processes. The overall findings of the three iterations can be summarized into four key components for STEM education in the context of preservice teacher education:

1. *Learning to teach integrative STEM*: Preservice teachers need to be actively involved in implementing STEM education as learners and be encouraged to reflect on the learning experience.
2. *Collaboration*: A way of working together where individuals respect and learn from each other reflecting the interdepence nature of STEM disiplines.
3. *Integration of STEM disciplines:* Solving real life problems through an interdisciplinary approach.
4. *Skills related to STEM disciplines:* Inquiry, modelling and engineering process skills are necessary for STEM education:

    (a) Inquiry: in the implementation of STEM education activities, preservice teachers used skill of inquiry to approach project-based learning task; to investigate scientific and mathematical roots to solve the problem.
    (b) Modelling: preservice teachers also needed to use modelling skills in order to handle the given real-life problem, mathematically model it as well as to represent their findings.
    (c) Engineering process: engineering may be the most overlooked discipline in STEM partly because there is not a common course in school curricula, and partly because its' process skills (design, test, redesign iterations) differs from scientific method (collect data, analyze and report results). Each activity in CLT-STEM required preservice teachers to use certain degree of engineering process skills.

These components are aligned with the dimensions proposed by Stohlmann, Moore, and Roehrig (2012), who listed their considerations for teaching integrated STEM in four folds; *support, teaching, efficacy*, and *materials*. Stohlmann, et al. (2012) developed their list after implementing a one-year integrated STEM education project with inservice science and mathematics teachers. In their list, support was defined as partnering with an institution, attending professional development and/or teacher collaboration; teaching included both lesson planning and classroom practices; efficacy is defined as the content and pedagogical knowledge of teachers that contributes to self-efficacy, commitment to STEM education and/or planning and organization; and finally, materials are resources (both technological and concrete) needed for STEM education. These considerations refer to the conditions needed for effective STEM teacher education, yet the components proposed by the authors in this chapter cover the method to be used provided by these conditions. For instance, the *support* and *materials* need to be given to teachers in an environment where they can *work collaboratively* and *learn how to teach integrated STEM*. On the other hand, self-efficacy did not emerge from the data of this study.

In conclusion, the CLT-STEM module, developed in three iterations, for chemistry and mathematics preservice teachers can be used as a model for implementing integrative STEM education and for fostering interdisciplinary collaboration. CLT-STEM can provide a framework for international audience of STEM education researchers and practitioners who wish to develop their own modules in other STEM disciplines.

Acknowledgement This project was funded by the Bogazici University Research Funds, Grant Number 9221.

References

Alan, B., Zengin, F. K., & Kececi, G. (2019). Using STEM applications for supporting integrated teaching knowledge of pre-service science teachers. *Journal of Baltic Science Education, 18*(2), 158–170.

Aslan-Tutak, F., Akaygun, S., & Tezsezen, S. (2017). Collaboratively learning to teach STEM: Change in participating pre-service teachers' awareness of STEM. *Hacettepe University Journal of Education, 32*(4), 794–816.

Baxter, J. A., Ruzicka, A., Beghetto, R. A., & Livelybrooks, D. (2014). Professional development strategically connecting mathematics and science: The impact on teachers' confidence and practice. *School Science and Mathematics, 114*(3), 102–113.

Berlin, D. F., & White, A. L. (2012). A longitudinal look at attitudes and perceptions related to the integration of mathematics, science, and technology education. *School Science and Mathematics, 112*(1), 20–30.

Bozkurt Altan, E., Yamak, H., & Buluş Kırıkkaya, E. B. (2016). Hizmetöncesi öğretmen eğitiminde FETEMM eğitimi uygulamaları: Tasarım temelli fen eğitimi [A Proposal of the STEM education for teacher training: Design based science education]. *Trakya Üniversitesi Eğitim Fakültesi Dergisi, 6*(2), 212–232.

Dugger, W. E. (2010). *Evolution of STEM in the United States*. 6th Biennial International Conference on Technology Education Research, Gold Coast, Queensland, Australia.

European Schoolnet. (2018). *Science, technology, engineering and mathematics education policies in Europe*. Scientix observatory report. October 2018, Brussels, Belgium: European Schoolnet.

Gökbayrak, S., & Karışan, D. (2017). STEM etkinliklerinin fen bilgisi öğretmen adaylarının bilimsel süreç becerilerine etkisi [An investigation of the effects of STEM based activities on preservice science teacher's science process skills]. *Batı Anadolu Eğitim Bilimleri Dergisi, 8*(2), 63–84.

Guzey, S. S., Tank, K., Wang, H.-H., Roehrig, G., & Moore, T. (2014). A high-quality professional development for teachers of grades 3-6 for implementing engineering into classrooms. *School Science and Mathematics, 114*(3), 139–153.

Honey, M., Pearson, G., & Schweingruber, A. (2014). *STEM integration in K-12 education: Status, prospects, and an agenda for research*. Washington, DC: National Academies Press.

Laboy-Rush, D. (2011). *Integrated STEM education through project-based learning*. Accessed on May 5, 2016 from http://www.rondout.k12.ny.us/common/pages/DisplayFile.aspx?itemId=16466975,

Ministry of National Education (MoNE). (2018). Fen bilimleri dersi öğretim programı (Science curriculum for grades 3–8). Accessed on March 9, 2019 from http://mufredat.meb.gov.tr/ProgramDetay.aspx?PID=325

Murphy, T. P., & Mancini-Samuelson, G. J. (2012). Graduating STEM competent and confident teachers: The creation of a STEM certificate for elementary education majors. *Journal of College Science Teaching, 42*(2), 18.

Nadelson, L. S., & Seifert, A. (2013). Perceptions, engagement, and practices of teachers seeking professional development in place-based integrated STEM. *Teacher Education and Practice, 26*(2), 242–265.

Nadelson, L. S., Seifert, A., Moll, A. J., & Coats, B. (2012). I-STEM summer institute: An integrated approach to teacher professional development in STEM. *Journal of STEM Education, 13*(2), 69–83.

National Research Council (NRC). (2010). *Rising above the gathering storm, revisited: Rapidly approaching category 5*. Washington, DC: The National Academies Press.

NGSS Lead States. (2013). *Next generation science standards: For states, by states*. Washington, DC: National Academies Press.

Perez, D., Gibson, J., Opsal, S. C. & Lynch, R. M. (2011). *Organizing an edible car contest a how to handbook*. Accessed on March 9, 2019 from https://atecentral.net/r35304/organizing_an_edible_car_contest

Project Oceanography. (1999). *Ocean color lesson I: Energy and color*. Accessed on March 9, 2019 from http://www.marine.usf.edu/pjocean/packets/sp99/s99u1le1.pdf

Science Buddies Staff. (2015). *Exploring nanotechnology: Fold, roll, & stack your way to super-strong materials*. Accessed on March 9, 2019 from http://www.sciencebuddies.org/science-fair-projects/project_ideas/MatlSci_p042.shtml

STEM Task Force Report. (2014). *Innovate: A blueprint for science, technology, engineering, and mathematics in California public education*. Dublin, CA: Californians Dedicated to Education Foundation.

Stohlmann, M., Moore, T. J., & Roehrig, G. H. (2012). Considerations for teaching integrated STEM education. *Journal of Pre-College Engineering Education Research, 2*(1), 4.

Thibaut, L., Ceuppens, S., De Loof, H., De Meester, J., Goovaerts, L., Struyf, A., et al. (2018). Integrated STEM education: A systematic review of instructional practices in secondary education. *European Journal of STEM Education, 3*(1), 2.

Tosmur-Bayazit, N., Akaygün, S., Demír, K., & Aslan-Tutak, F. (2018). An example of STEM teacher professional development: Exploration of edible cars activity from teacher education perspective. *Fen Bilimleri Öğretimi Dergisi, 6*(2), 213–232.

Vygotsky, L. (1978). Interaction between learning and development. *Readings on the Development of Children, 23*(3), 34–41.

Wang, F., & Hannafin, M. J. (2005). Design-based research and technology-enhanced learning environments. *Educational Technology Research and Development, 53*(4), 5–23.

Yıldırım, B., & Altun, Y. (2015). Investigating the effect of STEM education and engineering applications on science laboratory lectures. *El-Cezerî Journal of Science and Engineering*, 2(2), 28–40.

Yıldırım, B., & Cumhur, T. Ü. R. K. (2018). Pre-service primary school teachers' views about STEM education: An applied study. *Trakya University Journal of Education Faculty, 8*(2), 195–213.

Sevil Akaygun is an Associate Professor of Chemistry Education at Bogazici University, and a former science and chemistry teacher. Her research interests include visualizations in chemistry education, nanoscience education and STEM education. Dr. Akaygun has been coordinating national and international research projects and giving seminars and workshops to STEM teachers.

Fatma Aslan-Tutak is an Associate Professor of Mathematics Education at Bogazici University whose research interests are teacher education, mathematical knowledge for teaching, measuring mathematics teachers' knowledge, comparative teacher education studies, professional development to improve inservice teacher practices, classroom based assessment, integrated STEM education.

Part III
Critical Questions in STEM

Chapter 10
Public Portrayals of Indiana STEM Certified Schools

Steven Newman, Taukir Kahn, Meize Guo, Alex Gerber, Angela Burgess, and Valarie L. Akerson

10.1 Introduction

Education policy discussions increasingly focus on the urgent need for greater stress on integrated STEM instruction, where the interconnections and authentic, real-world relevance of these fields are emphasized (Hansen, 2014; National Center on Education and the Economy, 2007) to prepare students for the highly competitive twenty-first century, with new mindsets and skills (Breiner et al., 2012; Bybee, 2010; Morrison, 2006; Roberts & Cantu, 2012). Common instructional practices used within STEM education (project and inquiry based learning) have captured the attention of educators in K-12 classrooms as they enable students to relate their knowledge, skills, and beliefs across the STEM disciplines, (Breiner et al., 2012) thus promising more meaningful science and mathematics learning for K-12 students (Bicer et al., 2014). To achieve this, STEM funding for research and education has significantly increased over previous decades in the USA (Sanders, 2009).

Still today the meaning of STEM remains ambiguous and perhaps even considered political (Bybee, 2013). To others STEM can mean the integration of science, technology, engineering, and math curricula that more closely parallels the work of a real-life scientist or engineer. Others view STEM as a push for graduating more students in the science, technology, engineering, and mathematics fields so the United States can maintain its competitiveness and not fall behind emerging countries (Breiner, 2012). Within the classroom setting, STEM education can range from being situated around a single discipline (i.e. Science) or

S. Newman (✉) · T. Kahn · M. Guo · A. Gerber · A. Burgess · V. L. Akerson
Indiana University, Bloomington, IN, USA
e-mail: stevnewm@iu.edu

V. L. Akerson, G. A. Buck (eds.), *Critical Questions in STEM Education*, Contemporary Trends and Issues in Science Education 51,
https://doi.org/10.1007/978-3-030-57646-2_10

multi-disciplinary fields (i.e. Science and Math or STEM) (Bybee, 2013). Some school teachers consider technology as the bridge, connecting science, engineering, and math through technical equipment or software use in teaching (Radloff & Guzey, 2016). Computer science scholars claimed that computer science should be expanded as a part of K-12 STEM education (DeLyser et al., 2018) and computational thinking can be viewed as the core of all STEM disciplines (Henderson et al., 2007). An examination of the nature of each STEM discipline revealed the only characteristic that was really uniquely attributed to STEM was interdependence, even so this was deemed more the intention of STEM-that it describes the connections between and influences of individual disciplines on other STEM disciplines (Akerson et al., 2018).

Indiana responded to this call for STEM, spurred by an identified growing deficit in a highly-developed and technologically advanced workforce, citing that STEM disciplines are associated with skills that are the building blocks of Indiana's current and future economy (IDOE, 2019). This response is the driving force for increased STEM integration within Indiana's school systems and curriculum. The Indiana Department of Education (IDOE) defines STEM education as interdisciplinary literacy that seeks to integrate, in whole or in part, the four areas of science, technology, engineering, and mathematics into a comprehensive and coherent curriculum across content areas. A STEM classroom is defined as providing students opportunities to investigate and question the interrelated facets of the real world (Doe.in.gov, 2019). Collectively for this book, we hold the definition of STEM to be the meaningful interdependence among all disciplines of STEM. In other words, this includes all individual disciplines of STEM (science, technology, engineering and mathematics) in a way that is meaningful and showcases the interdependence of the fields.

Given the increased push for STEM incorporation in K-12 schools, we chose to investigate how Indiana STEM Schools publicly portrayed themselves after going through a rigorous certification program. To assess how schools are publicly portraying STEM, we chose to examine the following three questions: 1) How is STEM being portrayed across grade level bands (elementary, middle, high school) in STEM certified Schools? 2) What is being portrayed as STEM across grade level bands (elementary, middle, high school) in STEM certified schools? 3) Who is being portrayed as doing STEM across grade level bands (elementary, middle, high school) in STEM certified schools?

10.2 Literature Review

This section will examine some of the different ways STEM is perceived and how these perceptions impact how STEM is enacted. STEM will be considered as both an instructional approach and an educational outcome.

10.3 What Is STEM?

"STEM" has been an alluring word for the last several years. In 1990s, the National Science Foundation (NSF) introduced contemporary STEM as an acronym for science, technology, engineering, and mathematics. Bybee (2013) mentions that the ubiquitous and ambiguous uses of STEM in education have created much confusion. One of the possible reasons could be the lack of consensus on a definition of STEM. Without having a common definition for STEM, we have been developing and implementing various versions of STEM. Regarding STEM literacy, Bybee (2013) addressed that:

"In an effort to move from STEM as a slogan to an educational definition, I propose that STEM literacy refers to an individual's: 1) knowledge, attitudes, and skills to identify questions and problems in life situations, explain the natural and designed world, and draw evidence-based conclusions about STEM related-issues, 2) understanding of the characteristics features of STEM disciplines as forms of human knowledge, inquiry, and design, 3) awareness of how STEM disciplines shape our material, intellectual, and cultural environments, and 4) willingness to engage in STEM-related issues and with the ideas of science, technology, engineering, and mathematics as a constructive, concerned, and reflective citizen" (p. 70).

From the above components of STEM literacy, it is evident that students need to have real life experience where they can apply their knowledge and skill bases. This debate has still yet to been resolved. Are we considering STEM as a separate discipline or just an integrated curriculum approach? The idea of considering STEM as a separate discipline has been a puzzle to many educators. One of the most critical notions is that the STEM disciplines are all different ways of knowing and have different conventions for what constitutes data and evidence. Williams (2011) also acknowledged that STEM disciplines rely on different epistemological assumptions which could create a barrier toward the integrity of individual STEM subjects. STEM could be an integrated curriculum approach, but because we are dealing with different ways of knowing, true integration is never achieved; just an interdisciplinary connection. Wang et al. (2011) explained that interdisciplinary integration begins with a real-world problem which incorporates cross-curricular content with critical thinking, problem-solving skills, and knowledge in order to have a conclusion. Besides the interdisciplinary approach, English (2016) states that we can regard integration as: (1) multidisciplinary, where students learn contents separately under a common theme or (2) transdisciplinary, where students gather knowledge and skills from different disciplines in order to solve real world problem.

10.4 STEM as a Learning Outcome

STEM teachers are required to oversee their students' conceptions regarding the interconnection of all four STEM disciplines. Moore et al. (2014) emphasized the importance of interconnections in STEM curricula so that students can rectify the interconnections between different subject areas. There has not been much research done in what STEM means in terms of students' learning outcomes. After reviewing prior research, Honey et al. (2014) found no significant correlation between STEM integration and students' learning outcomes. In terms of STEM learning outcomes, Brooks (2016) addressed two factors, discipline-specific and broad educational outcomes. He specified discipline-specific outcomes as knowledge of individual disciplines-science, mathematics, engineering, and technology. Whereas, broad educational outcomes cover both the critical thinking and problem-solving skills among all the disciplines. Furthermore, these two types of learning outcome can come from either micro (school or university) level or macro (state or national) level.

10.5 Methods

This section will examine the methodological approach to data collection and analysis we followed to answer our three proposed research questions that relate to the public portrayal of Indiana STEM schools.

10.6 Research Context

Indiana schools seeking STEM certification must undergo a thorough and rigorous process prior to being granted certification by the Indiana Department of Education (IDOE). The process of transition from a traditional school to a STEM school requires forming connections within the community (i.e. businesses), incorporation of STEM at all grade levels, and adopting a new curriculum and associated instructional practices (Doe.in.gov, 2019). First schools need to complete a 22 page self-evaluation form that is submitted to the IDOE. Recommended adjustments by the IDOE need to be addressed by the school prior to completion of the in-depth, full application. The completed full application is submitted to a state STEM coordinator. Following application submission, a school site visit from the STEM review team, which consists of IDOE and local community representatives occurs. The onsite visit is compared with application materials to determine certification. Certification is only granted for those schools who demonstrate full STEM implementation through their application and meet the 85% school attributes mark including those deemed absolutely necessary as assessed via the IDOE STEM Implementation Rubric (Doe.in.gov, 2019). Schools that have attained certification must re-apply and demonstrate steps forward every 5 years. For the focus of this study, 42 schools attained STEM certification and based on the year certified and renewal time, are placed into 1 of 4 cohorts. Cohort 1 being the earliest (2015–2020) and cohort 4 the most recent (2018–2023) (Table 10.1).

In this study, we used content analysis, which is considered an appropriate research technique for making inferences from using text (Krippendorff, 2004;

Table 10.1 Grouping of schools analyzed by time of certification

Grade Level	Cohort 1	Cohort 2	Cohort 3	Cohort 4	Total
Elementary school	8	2	2	10	22
Middle school	2	3	4	6	15
High school	0	1	0	4	5
Total	10	6	6	20	42

White & Marsh, 2006). Since the research focused on how Indiana STEM certificated schools portrayed themselves to the public via their school websites, we followed the methods as used in Herring (2009), which applied traditional content analysis techniques to web content. McMillan (2000) summarized a procedural set of steps to perform content analysis, which we used in this study: (1) formulate research question, (2) select sample, (3) define categories for coding, (4) train coders to code the content, and (5) analyze and interpret the data.

Our stated research questions relating to the portrayal of STEM in STEM certified Indiana schools were: (1) How is STEM being portrayed across grade level bands, (2) What is being portrayed as STEM across grade band levels, and (3) Who is being portrayed as doing STEM across grade band levels? The Indiana Department of Education (IDOE) website lists all STEM certified schools in the state by cohort (60 initially). We used this published list to determine our sample size of relevant schools. We chose to exclude schools that were non-traditional elementary, middle or high schools (i.e. career centers/academies, technical schools, religion centered), which resulted in our final sample size (42). Since there were six of us, we then split into three teams of two and evenly divided the total number of schools to undergo coding for content analysis. We gathered each schools homepage URL link and compiled them in a Google Sheets spreadsheet to use as our starting point for content analysis. We defined three over-arching categories: 1) How, 2) What, and 3) Who is being portrayed doing STEM on each school's website. For our first round of content analysis, each of the three coding teams searched and coded their assigned schools website for information regarding the three previous categories that are based on our stated research questions. These codes were collaboratively combined in our shared Google Sheets spreadsheet and built upon.

After the first round of content analysis, we discovered that Twitter was primarily the main platform used by schools to publicly portray information in regards to STEM implementation. Therefore, each coding group performed a second round of content analysis for each schools Twitter account that was embedded on the school website. We checked all original posted tweets from July 2017 to November 2018 and coded for: (1) tweet content with the keywords: STEM, science, technology, engineering, math/mathematics, (2) pictures of activities attached to selected tweets, and (3) hashtags that were marked in the tweets (Fig. 10.1).

To insure congruence, we used inter-rater reliability by having two members in the same coding group met to compare data coding and interpretation of analyses.

Fig. 10.1 Sample STEM school's Twitter feed contents analyzed for coding. (**a**) Within the Twitter caption, Science is explicitly stated in the posting description. Therefore we can identify this as an instance of Science being portrayed. (**b**) While this appears to be Science, we did not count this as portraying Science, as it has not been explicitly labeled in the Twitter post. We did not make assumptions about what schools were portraying, only what was explicitly labeled

Through comparison of coding and interpretations, this allowed for resolution any differences in coding through further consultation of the data. We then all collectively met to discuss the final coding and interpretations. We also coded any additional social media platforms for each school, such as Facebook, a YouTube channel, and school newsletters if the content appeared to be absent from Twitter.

10.7 Results

In this section we report our findings about how STEM certified schools are portraying STEM. We have separated our findings by specific research question and grade band. We begin with "how" STEM is being portrayed, then "what" is being portrayed as STEM and finally "who" STEM is being portrayed doing STEM between each grade band.

10.8 How Is STEM Being Portrayed across STEM Certified Grade Bands?

For the eight elementary schools that acknowledge STEM certification, six only had a respective STEM logo displayed on their respective homepages (Table 10.2). Eleven of the 22 schools mentioned to varying degrees they were STEM certified within their "school information" sections. Twenty-one of 22 schools had twitter accounts of which 19 portrayed varying degrees of activities tagged as STEM (Table 10.2). Facebook and YouTube were used intermittently on an individual school basis. Interestingly, of the 22 elementary schools, 12 had some sort of STEM specialist (teacher, coach, or trainer) listed on their school websites (Table 10.2). Two schools have incorporated the term STEM into their school names (i.e. Washington STEM Academy and Brooklyn STEM academy).

Table 10.2 School website coding instances for "How is STEM being portrayed across STEM certified grade bands?"

	Elementary	Middle	High
Recognized STEM Certification[a]	8	5	5
Had social media	21	15	5
Twitter[b]	(19)21	(9)15	(2)5
FaceBook[b]	(7)16	(0)2	(0)0
YouTube[b]	(1)2	(5)5	(0)2
STEM specialist	12	0	0
Total schools per Grade band	22	15	5

[a]The schools websites visually show some form of STEM certification endorsement to the public, to acknowledge the rigorous process they under went to achieve certification from the Indiana Department of Education (IDOE)

[b](Showcased school identified STEM activities) total number of schools with social media type

Five of the 15 middle schools use their school website homepage to communicate about STEM (Table 10.2). The home page representation always included the IDOE certification logo. Some schools promoted their STEM certification in a prominent place on their website homepage, for example in the banner picture gallery, while others have a smaller STEM certification logo in less conspicuous locations such as at the bottom of the homepage. It should be noted that one school included the term STEM in the school name (i.e. Maple Crest STEM Middle School). Another common way that STEM is being portrayed is via the school information section on their websites. Seven of the 15 schools directly referred to STEM in various locations on their sites. For example, some schools included a "message from the principal" section where STEM was mentioned, others had an "about" section that included STEM information, two even provided curriculum breakdowns that included explicit STEM references. Many schools were utilizing social media platforms as a tool to communicate with parents, students and the community about school related activities and initiatives. Nine schools in total used at least one social media platform to communicate STEM ideas. Twitter is the most commonly utilized platform, with YouTube and Facebook used less frequently (Table 10.2). YouTube videos depicting STEM were often embedded into pages on the school website. Some schools which had no STEM presence on their websites seemed to utilize social media platforms as their main source of communication. Only two schools appear to use newsletters as a source of communication, and only one of those schools use the newsletter to communicate about STEM.

Interestingly, cohort 1 middle schools use their school websites to demonstrate how STEM is being integrated into the curriculum. This finding is not observed in schools from other cohorts. Cohort 4 middle schools, those most recently certified, exhibited far less STEM communication than the other cohorts across all media platforms despite there being an increase in number of schools being certified.

All STEM certified high schools shared on their main websites that they are STEM Certified schools (Table 10.2). One of these high schools had a STEM Academy embedded within the high school, as one of its several academies. They provided a general statement and a STEM logo, along with a brief statement about being a STEM certified school. Two of the schools had a twitter account, and occasionally produced tweets to describe STEM events, but these are not too frequent (Table 10.2). One school website announced a STEM-themed essay contest. Two schools had a STEM newsletter linked to the website. One school described how they teach science and mathematics, but did not appear to integrate the STEM disciplines. Another school described an integrated program that "works across the disciplines, including math, science, reading, social studies, and other subjects" but did not describe how their curriculum does this type of teaching. A third school described the importance of STEM in terms of designing innovative products and supporting the future, incorporating arts within its curriculum. The fourth school described STEM as being hands-on activities that required students to use their creativity to learn more about science, technology, engineering, and mathematics. The final high school provided instruction within the separate disciplines and did not seem to integrate across the disciplines. At this school they hold events, such as robotics challenges, that incorporate their STEM program.

Because four of the high schools were in their first year of being a STEM-certified school, there may have been fewer tweets and sharing of events. For example, one high school shared photos of students building a light box and also building robots. One of the cohort 4 schools tweeted about STEM careers for women.

10.9 What Is Being Portrayed as STEM across Grade Level Bands?

Twitter feed content analysis revealed collectively what is being portrayed as STEM in Indiana STEM certified elementary schools. A noteworthy finding of elementary schools' was that 14 of the 22 schools' Twitter feeds showcased a robotics team in competition or kids working with the robots themselves as part of their STEM initiative (Table 10.3). Fifteen of 22 schools associated STEM challenges or projects involving building specifically labelled as engineering or engineering design (Table 10.3). Four schools were found to have STEM designated classrooms for which the various grades levels within the school were shown working on STEM labelled activities and three schools also had their own makerspaces (Table 10.3). Interestingly, only four schools portrayed teachers participating in STEM related professional development (Table 10.3). Unique examples of STEM integration into schools included a robotics room and makerspace (i.e. Indian Creek), and incorporation of STEM kits (i.e. Indian Creek and Stout Field). To enhance their schools' STEM in the classroom, four schools showcased to varying degrees teacher professional development (i.e. STEM committee, technology incorporation, science).

Twitter feed content analysis allowed us to examine to what extent each school portrayed what they perceive as STEM by what they specifically tagged. As STEM certified schools, one would think the use of STEM would be at the forefront of the school's public image, however it was found to vary across three levels. We classified as five schools as level 1 certified STEM schools. Their Twitter feeds were either district wide accounts or simply showcased very minimal acknowledgement of STEM. Four of the 5 schools were from the most recent cohort group 4, whereas

Table 10.3 Twitter feed coding instances for "What is being portrayed as STEM across grade level bands?"

	Elementary	Middle	High
Robotics team	14	4	2
STEM challenges/projects	15	4	1
Professional development	4	2	0
STEM classroom	4	2	0
Makerspace	3	1	0
Total schools per grade band	22	15	5

one was from cohort 1. Level 2 STEM classified schools ranged in STEM representation of the individual STEM disciplines being tagged specifically to the identification of activities as STEM. Within this level, engineering was more frequently tagged in conjunction with STEM, not on its own. Schools within level 2 consisted of schools from all 4 cohorts of STEM certification. Level 3 STEM certified schools had representation from each of the 4 cohort certification bands as well. These schools portrayed a wide range of diverse STEM discipline activities or projects ongoing at the school ranging from simple to complex (i.e. Washington, playing with water to recreating the events of the titanic, respectively). One specific discipline was not showcased or focused on more than the others and STEM labelled activities showcased students participating in multi-discipline activities collectively as opposed to level 2 schools which labelled single disciplines as STEM throughout their Twitter feed.

STEM is being portrayed in various ways across the middle school grades. Only four schools showcased a robotics team and one school showed a makerspace (Table 10.3). Within the middle schools, far fewer where labelling STEM activities as challenges or projects compared to elementary schools (Table 10.3). Two schools had STEM designated classrooms and showed teachers in STEM related professional development (Table 10.3).

By analyzing curricula that were posted on two cohort 1 school websites it can be observed that STEM is being used as an end-of-semester project which attempts to integrate concepts from science and math. Despite the STEM acronym being utilized, some schools clearly represent STEM as inclusive of disciplines including English, Language Arts and Social Studies while technology and engineering are not represented explicitly in their curricula. Interestingly, schools were unlikely to portray separate S, T, E and M classes as STEM, there had to be some integration across at least two disciplines in order for an activity or lesson to be cited as STEM. Digital technology use appeared to be an indicator as to whether an activity and or lesson will be cited as STEM in middle school social media posts. For example, a cohort 3 school would not cite an animal dissection activity as STEM, yet the use of Google Expedition VR goggles to explore cell organelles was cited as STEM. As well as the apparent prerequisite of some type of integration between disciplines, STEM was cited most frequently when referring to a model building/making activity that typically takes place within science lessons. There is only one instance where STEM was cited within a mathematics lesson, and this was when digital technology was a conspicuous tool in the classroom. However, it is interesting to note that technology-based activities such as coding and programming were not typically described as STEM activities. Another common theme of STEM portrayal was STEM as twenty-first century skills such as critical thinking, collaboration, innovation and communication. In some cases, STEM is being portrayed as resources that are available to schools. For example, a cohort 3 school used social media to communicate about their technology enabled STEM classrooms, and another school promotes its 1:1 digital tablet ratio as a STEM initiative.

Invariably, middle schools utilize general STEM activities. One important exception to this is a single middle school that promoted agri-STEM, STEM that focused specifically on agricultural issues that may be more relevant to their rural student population.

STEM is being portrayed in various ways across the high schools examined. Only two of the five portrayed a robotics team, with no schools portraying a designated STEM classroom or makerspace (Table 10.3). Only one school showed STEM being done as a challenge (Table 10.3). No high schools twitter feed content showed any STEM related professional development (Table 10.3).

STEM is portrayed as being taught in different ways in all the STEM certified high schools in terms of the number of subjects they cover. It appeared that all the schools teach four subjects (S,T,E,M) separately even though they call it STEM. Across the five high schools "science" and "mathematics" dominated in most of the schools' portrayal of STEM instruction in the classroom. For example, two schools from cohort 4 relied on science and mathematics teaching in the name of STEM. Furthermore, both the technology and the engineering parts were somehow not visible in the curriculum of most of the high schools. One school incorporated science, mathematics, and information technology as separate subjects in their curriculum. Another school put great emphasis on both the technology and the engineering subjects. For example, under the name of the engineering and the technology courses, they have different pathways or sub-disciplines: engineering, robotics, architecture, manufacturing, and construction pathways. Each of the pathways has individual activities or projects with the fixed goals and objectives. For example, PLTW (Project Lead the Way) under the engineering pathway helps students develop the skills needed to succeed in our global economy.

10.10 Who Is Portrayed as Doing STEM across Grade Level Bands?

For elementary schools the majority depicted students participating in STEM activities, however, there were also pictures that focused on professional development (PD) in STEM for teachers (Table 10.4). Several of the professional development sessions, as well as some of the activities for the students, were led by a STEM specialist. Of the twenty-two schools only 12 had some sort of STEM specialist in house, whereas other schools brought in a STEM specialist from outside the school (Table 10.2). For example, Stout Field Elementary had a STEM specialist come to their school to teach students about biomes. While STEM specialists occasionally came to the schools, most of the STEM instruction was being carried out by the regular classroom teachers at the schools. In some instances, twitter posts portrayed external companies/people coming to a school and was tagged as STEM (Table 10.4). For example, one school had their local weather station crew come by or representatives from the local zoo come to school. Only a few elementary schools publicized STEM social events/nights put on by their schools (Table 10.4). Overall the pictures

Table 10.4 Twitter feed coding instances for "Who is being portrayed as doing STEM across grade level bands?"

	Elementary	Middle	High
Students	20	9	3
Teachers	9	5	1
Parents	2	1	0
External/outsourced people	8	3	1
STEM socials/nights	7	1	0
Total schools per grade band	22	15	5

posted on the school websites and the school twitter feeds were focused on students completing what the school deems to be a STEM activity.

There are a variety of people who are portrayed as doing STEM by the STEM certified middle schools. Typically students were portrayed as engaging in STEM learning via participation in STEM based classroom activities (Table 10.4). Students were usually depicted as working in groups in dynamic classroom environments. Teachers are also portrayed as "doing STEM" in PD contexts (Table 10.4) and were often centered on technology integration into discipline specific classrooms, but in at least one instance, teachers from a cohort 3 school utilized local STEM professional industry experts' knowledge in their professional development. STEM professionals were invited into schools and portrayed as "doers of STEM" by the schools especially in instances where STEM careers where the emphasis of an activity (Table 10.4). Finally, in one instance families and the wider community were portrayed as engaging in STEM via a STEM night event (Table 10.4).

Primarily students and in one instance teachers and STEM professionals are being portrayed doing STEM for high schools (Table 10.4). Each of the five high schools' students were involved in STEM programs as per described through their social media activities. For example, four schools depict their students' active involvement such as robotics (Table 10.3), STEM workshops, and STEM essay competitions. Women's participation in STEM programs are highly notable in all the high schools. One high school included a program called "Women in tech careers". Teachers were shown participating occasionally in the role of facilitator (Table 10.4). Either a science or mathematics teacher conducts different STEM related activities and also teach technology and engineering courses. In one school, there were just science and mathematics teachers to teach all four disciplines. There is no parental involvement visible in any of the STEM programs. One school regularly invited STEM professionals from outside the school to organize STEM essay competitions under their STEM initiative program (Table 10.4).

10.11 Discussion

In this section we will discuss the implications of our results about how STEM certified schools are portraying STEM. We separated our discussion by each of the three specific research questions.

10.12 How STEM Is Portrayed across Grade Bands?

Across all three grade bands of schools within Indiana who are now STEM certified, not all prominently display the results of this rigorous school endeavor to the public on their school homepages. One would think that the effort and potential costs associated to the district to transition traditional school(s) to STEM certified school(s) would warrant more public advertising to gain interest in new student enrollment at a "STEM" school over a neighboring traditional school. Many of the school websites portrayed a school Twitter account to provide insight into the activities of the school ranging from multiple posts per day to few posts a month. The sources of the Twitter posts varied from discipline specific teachers to school dedicated accounts of which the control of was not apparent. It was not clear if part of STEM certification process required demonstration of how STEM was being implemented within the classrooms or if these were all personal choices by the respective teachers or schools. However personal teacher accounts warranted higher frequencies of posts compared to district accounts that were convoluted with all ranges of school activities, not just STEM related events. The portrayal based on individual teachers is likely spurred on by their personal excitement and enthusiasm for engaging their students in new STEM activities for their subjects taught.

Only elementary schools were identified as having some form of STEM specialists listed on their respective school websites (Table 10.2). Furthermore, elementary twitter feeds revealed that if a school did not have a STEM specialist listed they would bring one in on various days. Given the content mastery required of elementary teachers as more likely to be generalists compared to middle and high school teachers who are more discipline specific and isolated, is it easier to become a "STEM specialist"? It has been found that between grade bands elementary teachers were more likely to believe that they were already integrating STEM subjects, seemingly because separate subject classes are not as commonplace, especially in the early grades. In middle and high school, the extent of integration was mixed, but many teachers had experimented with integration in some meaningful way, with some projects involving coordination across subjects (Shernoff et al., 2017). Given the surface level of content and openness of the classroom setting, elementary provides a more amenable environment for specializations in curriculum development and implementation by those deemed "STEM specialists". A survey assessing the needs of teachers in regards to STEM education found that school administrators are interested in hiring STEM teachers, there is however a lack of candidates due to a shortage of truly integrated STEM pre-service education. Ideal STEM teachers would not be intimidated by mathematics and science, but rather would have a passion for and a solid content expertise in both—along with proficiency in related disciplines. Good candidates would also have a comfort level with exploring uncharted territory, trying new things, adapting, and the inevitable failure that comes with experimentation (Shirnoff et al., 2017).

10.13 What Is Being Portrayed as STEM across Grade Bands?

STEM appears to be ambiguous and not consistent across grade levels, or across schools within grade level bands. For elementary schools STEM subjects appear more integrated, whereas STEM subjects appear increasingly separated from middle to high school. Within elementary schools the majority of schools subsidized engineering with robotics clubs and/or STEM challenges that include engineering. Engineering was rarely tagged individually on its own. A meta-analysis of robotics use in association with schools curricula identified the use of robotics as a predominantly extracurricular activity (57% of studies) in relation to the curriculum remains prevalent (Benitti & Spolaôr, 2017). Robotics is likely used to stand in for engineering as they come in package sets that can be readily setup to use and requires building. The emphasis on an extracurricular activity is that in-service teachers are not trained or required to take engineering classes during their pre-service years at least at Indiana University. Teachers often cite the lack of content knowledge and understanding as the major challenges to integrating STEM into their classroom (Shirnoff et al., 2017), so this discomfort with the processes of engineering is shifted to after school. Which could continue to propagate the prevalence of mathematics and engineering as underrepresented in studies claiming to address STEM education (English, 2016).

For middle school STEM seems to additionally include English, language arts and social studies, rather than an increased focus on technology and engineering. For instance, an animal dissection activity was not cited as STEM, but a Google expedition on cell organelles was labelled as STEM. A common theme observed in middle school was STEM meant integration of two disciplines, and use of technology increased the likelihood of an activity being tagged as STEM. Incorporation of STEM was portrayed as end of semester activities that the students would be involved in. Since middle school teachers are a hybrid between content generalists and specialists, the ability to incorporate multidisciplinary STEM becomes more challenging while trying to balance the demands of standardized testing. The appearance of social studies and language arts begins to make sense when one considers what disciplines various middle school teachers may be required to cover. So in an effort to include STEM, they merge or combine two or more of their respective disciplines.

For STEM certified high schools, each STEM discipline is taught separately despite being labelled as STEM. Math and science were the disciplines that dominated STEM instruction. Integration is mentioned as happening, but not represented as tagged via student activities on the publicly available school websites and social media. Where integration was more likely occurring was in Twitter tagged pictures of after school competitions. One trend apparent that differentiates high school portrayal more so than middle and elementary schools was the focus on future and potential career avenues for students post-high school. As students' progress from elementary to high school does what is portrayed or considered STEM shift from interactive, collaborative activities in the classroom to a more specific career application and focus beginning in middle school and continuing throughout high school.

10.14 Who Is Being Portrayed as Doing STEM across Grade Bands?

Across all age bands, most commonly portrayed were the faces of the schools, the students participating in classroom activities, usually in groups, either labelled as STEM or its individual disciplines. Rarely would the teacher be captured within these public postings. Occasionally, when extracurricular STEM nights where hosted, students' parents would be shown interacting with the students engaged in STEM activities. When teachers were tagged as engaged in STEM it would be without the students, in professional development (PD) settings. Not every schools social media indicated professional development was being done across age bands. Yet, when asked to incorporate STEM many teachers have discussed the need for more teacher PD. It is indicated that in order to be effective, teachers need additional PD on how to incorporate and connect different STEM disciplines and on technology integration (Shernoff et al., 2017). As STEM appears within the classrooms it presents teachers the challenge of adapting to a new teaching strategy that showcases the interdisciplinary interactions of STEM disciplines. Some commonly tagged images associated with elementary teacher PD showed teachers learning how to incorporate new math related technologies into their classrooms. Without continued in-service PD and pre-service teacher training focused around the STEM disciplines the degree of the teacher's ability to integrate across disciplines will be limited as they will not know how to make the critical and meaningful connections between the disciplines. Once the teachers see the value of integration, then teachers will be more willing to makes those connections for their students (Shernoff et al., 2017). A shift in how STEM is labelled on social media and school websites may change from individual specific discipline classes to collaboration between teachers and disciplines being portrayed.

10.15 Conclusion

Even within the state of Indiana, that has a rigorous process by which schools become STEM certified, the portrayal of STEM across and within grade bands is highly variable. While we only gained insight into the portrayal of what schools are tagging as STEM on social and digital media, from what we can see, it appears that elementary schools try to integrate the disciplines while middle and high school grow increasingly separate discipline-wise. This could be due to the fact that elementary schools are designed in a way where one teacher is responsible for all subjects, as opposed to high school where each teacher has their own subject. The generalist feature of elementary schools and lower level content knowledge required may enable easier integration of STEM. One interesting future direction is the extent to which future STEM teachers will need expertise in the standards and key concepts of all other STEM subjects, or whether there will be some consolidated set of principles considered most important for the STEM teacher of the future to master.

References

Akerson, V. L., Burgess, A., Gerber, A., Guo, M., Khan, T. A., & Newman, S. (2018). Disentangling the meaning of STEM: Implications for science education and science teacher education. *Journal of Science Teacher Education, 29*(1), 1–8.

Benitti, F. B. V., & Spolaôr, N. (2017). How have robots supported stem teaching? In *Robotics in STEM education* (pp. 103–129). Cham, Switzerland: Springer.

Bicer, A., Navruz, B., Capraro, R. M., & Capraro, M. M. (2014). STEM schools vs. non-STEM schools: Comparing students' mathematics state based test performance. *International Journal of Global Education, 3*, 3.

Breiner, J. M., Johnson, C. C., Harkness, S. S., & Koehler, C. M. (2012). What is STEM? A discussion about conceptions of STEM in education and partnerships. *School Science and Mathematics, 112*, 3–11. https://doi.org/10.1111/j.1949-8594.2011.00109.x

Brooks, C. L. (2016). *Understanding STEM learning outcomes using a Phenomenographic approach* (Unpublished doctoral dissertation). University of Massachusetts Amherst.

Bybee, R. W. (2010). Advancing STEM education: A 2020 vision. *Technology & Engineering Teacher, 70*(1), 30–35. https://doi.org/10.1017/CBO9781107415324.004

Bybee, R. W. (2013). *The case for STEM education: Challenges and opportunities*. Arlington, VA: National Science Teachers Association Press.

DeLyser, L., Goode, J., Guzdial, M., Kafai, Y., & Yadav, A. (2018). *Priming the computer science teacher pump: Integrating computer science education into schools of education*. New York: CSforAll.

Doe.in.gov. (2019). *Indiana's STEM Certified Schools | IDOE*. [online] Available at: https://www.doe.in.gov/wf-stem/indianas-stem-certified-schools. Accessed 14 Feb 2019.

English, L. D. (2016). STEM education K-12: Perspectives on integration. *International Journal of STEM Education, 3*(1), 3.

Hansen, M. (2014). Characteristics of schools successful in STEM: Evidence from two states' longitudinal data. *The Journal of Educational Research, 107*(5), 374–391.

Henderson, P. B., Cortina, T. J., & Wing, J. M. (2007). Computational thinking. *ACM SIGCSE Bulletin, 39*(1), 195–196.

Herring, S. C. (2009). Web content analysis: Expanding the paradigm. In *International handbook of internet research* (pp. 233–249). Dordrecht: Springer.
Honey, M., Pearson, G., & Schweingruber, A. (2014). *STEM integration in K-12 education: Status, prospects, and an agenda for research*. Washington: National Academies Press.
Krippendorff, K. (2004). Reliability in content analysis: Some common misconceptions and recommendations. *Human Communication Research, 30*(3), 411–433.
McMillan, S. J. (2000). The microscope and the moving target: The challenge of applying content analysis to the world wide web. *Journalism & Mass Communication Quarterly, 77*(1), 80–98.
Moore, T. J., Stohlmann, M. S., Wang, H., Tank, K. M., Glancy, A. W., & Roehrig, G. H. (2014). Implementation and integration of engineering in K-12 STEM education. In S. Purzer, J. Strobel, & M. Cardella (Eds.), *Engineering in pre-college settings: Research into practice* (pp. 35–60). West Lafayette, IN: Purdue University Press.
Morrison, J. S. (2006). *Attributes of STEM education: The students, the academy, the classroom*. TIES STEM education monograph series. Retrieved From https://goo.gl/J4CiUq.
National Center on Education and the Economy. (2007). *The report of the new commission on the skills of the American workforce*. San Francisco, CA: Jossey-Bass.
Radloff, J., & Guzey, S. (2016). Investigating preservice STEM teacher conceptions of STEM education. *Journal of Science Education and Technology, 25*(5), 759–774.
Roberts, A., & Cantu, D. (2012). Applying STEM instructional strategies to design and technology curriculum. In *PATT 26 conference; technology education in the 21st century* (Vol. 073, pp. 111–118). Stockholm, Sweden: Linköping University electronic press.
Sanders, M. (2009). STEM, STEM education, STEMmania, *The Technology Teacher,* Dec./Jan.-2009, 20-26.
Shernoff, D. J., Sinha, S., Bressler, D. M., & Ginsburg, L. (2017a). Assessing teacher education and professional development needs for the implementation of integrated approaches to STEM education. *International Journal of STEM Education, 4*(1), 13.
Wang, H., Moore, T., Roehrig, G., & Park, M. (2011). STEM integration: Teacher perceptions and practice. *Journal of Pre-College Engineering Education Research, 1*(2), 1–13.
White, M. D., & Marsh, E. E. (2006). Content analysis: A flexible methodology. *Library Trends, 55*(1), 22–45.
Williams, J. (2011a). STEM education: Proceed with caution. *Design and Technology Education, 16*(1), 26–35.

Steven Newman is a PhD student in Science Education at Indiana University. He transitioned from a background in scientific bench research to further education in teaching science at the college level. His research interests are on examining the interaction of co-teaching on teacher knowledge and student-teacher interactions.

Taukir Kahn is a third year PhD student of Science Education at Indiana University. His research area focuses on pre-service elementary teachers' professional knowledge bases in science. Additionally, he is interested in developing new instruments to portray science teachers pedagogical content knowledge.

Meize Guo is a Doctoral Candidate in Instructional Systems Technology and minored in Science Education at Indiana University. Her research focuses on technology integration and computer science education, especially on teacher education and professional development. Currently, she is researching elementary STEM teachers' conception and practice of teaching computer science.

Alex Gerber is a Doctoral Candidate in Science Education at Indiana University. He has experience teaching in a variety of informal contexts as well as college courses in the school of education at Indiana University. His research focuses on informal science education and, more recently, the use of representations in science teaching.

Angela Burgess is a Doctoral Student of Science Education at Indiana University and an informal environmental educator. Her research focuses on environmental literacy, environmental education practices and participant outcomes.

Valarie L Akerson is a Professor of Science Education at Indiana University, and a former elementary teacher. Her research focuses on preservice and inservice elementary teachers' ideas about Nature of Science, as well as their teaching practices. She is a Past-President of the Association for Science Teacher Education and a Past-President for NARST: A worldwide organization for improving science teaching and learning through research.

Chapter 11
Current Praxis and Conceptualization of STEM Education: A Call for Greater Clarity in Integrated Curriculum Development

Christopher M. Sgro, Trisha Bobowski, and Alandeom W. Oliveira

11.1 Statement of Objectives

Over the past 30 years, a drive towards integrated STEM has permeated the rhetoric of the academic field of education as well as educational policy (Honey, Pearson, & Schweingruber, 2014; Hudson, 2015). A cursory glance at the United States Department of Education (n.d.) website yields an entire subpage for the promotion of STEM education for global leadership, including substantial grants to produce stronger STEM-based instruction. This political and academic push has led to a vast and diverse amount of publications involving integrated STEM education practices at the K-12 school level. However, consensus is yet to be achieved as to what STEM education exactly entails. This problematic state of affair is highlighted by Kysilka (1998), who emphasizes that

> At the moment it seems that integration means whatever someone decides it means, as long as there is a 'connection' between previously separated content areas and/or skill areas. Before any teachers or administrators can successfully plan for integrated curriculum, a much clearer concept of what is meant by integration needs to be understood (p. 198).

Recent research has revealed considerable confusion, misunderstanding, and variance in how educators view STEM teaching and learning (Breiner, Harkness, Johnson, & Koehler, 2012; Radloff & Guzzey, 2016). There is also general ambiguity surrounding what constitutes various levels of integration, what the boundaries for each discipline are, and even how many disciplines are integrated (Sanders, 2009; English, 2016). Becker and Park (2011) report that, out of twenty-eight empirical studies of STEM education, seventeen integrated across only two of the four disciplines and only one study actually integrated across all four disciplines.

C. M. Sgro (✉) · T. Bobowski · A. W. Oliveira
Department of Educational Theory and Practice, State University of New York, Albany, NY, USA
e-mail: csgro@albany.edu

V. L. Akerson, G. A. Buck (eds.), *Critical Questions in STEM Education*, Contemporary Trends and Issues in Science Education 51,
https://doi.org/10.1007/978-3-030-57646-2_11

The existing variation in what is being published and purported as integrated STEM curricula suggests a general need for greater clarity about not only what constitutes STEM education, but how educators as a whole conceptualize STEM and the process of integration.

In an effort to add clarity to educational efforts in this area, this chapter examines current praxis and conceptualization of STEM among K-12 practitioners. More specifically, we conduct a comprehensive and systematic examination of integrated STEM curriculum recently featured in practitioner journals across these four disciplinary areas. Our analytical effort takes into account both the *intended curriculum* (what is explicitly/implicitly envisioned as STEM by practitioners of various school subjects) and the *operationalized written curriculum* (how their vision is operationalized as curricular materials to be put into action by teachers) (van den Akker, 2004).

11.2 Our Position on Integrated STEM Education

In an effort to better understand our findings as well as our drawn conclusions, it is important to understand how we position ourselves with regard to STEM, STEM education, and integration. As will be readily apparent in this chapter, we view STEM and STEM education as nebulous terms that vary based on individual stakeholders and their goals. It is our opinion that STEM is a "meta-discipline" (Kennedy & Odell, 2014) that integrates knowledge and skills from each of the four disciplines–science, technology, engineering, and mathematics–transforming that knowledge so that "the result is more than the sum of these parts" (Gibbons, 1979). Furthermore, through the process of integration, STEM uses the skills and knowledge of each of these disciplines in a way that addresses complex, real-world problems. It is our belief that this view of STEM strongly aligns with the collective definition that is woven throughout this book, where STEM showcases a meaningful interdependence among all of the individual disciplines of science, technology, engineering, and mathematics. Moreover, we believe that our examination of current practitioner work in STEM education supports this vision of STEM, as we investigate the variations in how educators operationalize STEM to highlight interdependence between individual disciplines through the process of integration.

Based on our position of what constitutes STEM, our views of STEM education surround the construction of lessons that integrate knowledge and skills from all four of these disciplines to solve complex, real world problems in a manner that is both engaging and student-centered. We believe that true integration involves the inclusion of *all* components, so an integrated STEM lesson must include science, technology, engineering, and mathematics knowledge or skills in some way where explicit connections are made to the underlying disciplines. While we do not believe that all components must be equally represented in every lesson, we do believe that there must be enough of a balance to be transformative, making it difficult to isolate the specific disciplinary contributions within the lesson. It is our belief that it is in this transformation, where the barriers between disciplines begin to erode, that deeper and more authentic learning can occur.

11.3 Theoretical Framework

The present examination of the practitioner literature on STEM education is informed by theories of integration from the field of curriculum studies. Of particular relevance to our work is Gibbons' (1979) theory of *integration by metamorphosis*, which emphasizes the transformative nature of integration. As Gibbons (1979) writes, "to integrate, logically speaking, is to unify so that the result is more than the sum of these parts" (p. 321); the ultimate result is transformation of knowledge. Further, for integration between two disciplines with different knowledge and ways of knowing to occur, one domain must be used to modify knowledge used by another. In this process, there are two domains: the domain of enquiry and the instrumental domain. The domain of enquiry acts as the discipline in which knowledge will be modified. The instrumental domain uses its knowledge, skills, and ontological views to modify knowledge found in the enquiry. Once modified, the newly integrated knowledge is then converted into a manner that is usable and applicable to the domain of enquiry. The quadripartite nature of STEM integration underscores the possibility of there being additional domains and the culmination of a "meta-discipline" (Kennedy & Odell, 2014) that is grounded in its four representative domains and takes its own form from the overlap of these domains and the removal of traditional boundaries among school subjects.

The removal of boundaries between disciplines occurs during higher levels of integration. Recently, Vasquez (2014) further illuminated this point with the inclined plane of STEM integration, a conceptual framework that designates degrees of integration within STEM instruction. At the lower level, there is disciplinary work, in which the knowledge and skills learned as part of the instruction serve the single discipline. When knowledge and skills from outside disciplines are introduced by instructors to demonstrate connection, a multidisciplinary approach takes place. However, when two or more disciplines are taught together to tightly link the skills and knowledge, a deeper level of interdisciplinary work is established. Finally, at the highest level of transdisciplinary integration, students solve real-world contextual problems using the knowledge and skills from multiple disciplines. At these highest levels, transdisciplinary work is linked with problem-based instruction (PBL).

The concept of transdisciplinary integration is similar to the models of curriculum integration posed by Moore, Stohlmann, Wang, Tank, Glancy, and Roehrig (2014), in which *content integration* is distinguished from *context integration*. In the model of content integration, a lesson is constructed with the goal of "merging multiple STEM content areas into a single curricular activity in order to consider the 'big ideas' from the multiple content areas" (p. 39). The overlap of these big ideas is where transdisciplinary integration exists, as the boundaries between disciplines begin to disappear. In contrast, context integration is when a lesson "primarily focuses on the content of one discipline and uses contexts from others to make the content more relevant" (p. 40). In short, context integration is when the goals and objectives of one discipline are highlighted over another.

The above curricular models of integration highlight the issues that currently affect STEM integration. In order to approach transdisciplinary integration, where traditional boundaries disappear, a real-world contextual problem is posed, and the students use knowledge and skills to solve it, regardless of if they are from science, technology, mathematics, or engineering. If the problem exists within a context that highlights one discipline's skills or knowledge, it has the potential to diminish the epistemological and ontological skills that the other disciplines can provide. This can result in a learning experience that is imbalanced. It is our belief that such an imbalance can lead to a fragmented view of problem-solving, diluting the authenticity of the inquiry itself.

11.4 Methodology

11.4.1 Data Collection

Consistent with research traditions in the scholarly field of curriculum studies, we conducted a comparative curriculum analysis. As defined by Egan (1978), "curriculum inquiry is educational inquiry; both properly address the *what* and *how* questions together and deal with all ramifications of trying to answer, 'What should children learn, in what sequences, and by what methods?'" (p. 70). Curriculum scholarship examines the type of curricular spaces created by educators, how curricular boundaries are made and unmade, and how such (re)configurations contribute to curricular reconstruction (Pinnar, 2011).

We focused our search on secondary-level articles published in STEM practitioner journals since 2010. The articles were found using Google Scholar, using key word searches for "STEM," "STEM Education," and "STEM lessons." These searches yielded several practitioner journals that regularly published STEM education articles, such as *Science Scope*, *The Science Teacher*, *Technology & Engineering Teacher*, and *Mathematics Teacher*, among others. We chose to focus on these journals because of their regular publication of integrated STEM lessons and their reputation for publishing articles written by practicing teachers in the STEM fields. By focusing on these journals specifically, we were able to keep the survey broad yet manageable. However, we do recognize that specific journals often impose their own criteria for published lessons, which could limit the generalizability of our findings beyond our survey scope.

Once we selected our practitioner journals, we combed the published STEM lessons since 2010, giving priority to lessons that were more recent. We selected lessons based upon their integration of multiple disciplines as well as their novel and useful nature. In collecting the articles, we aimed for six to ten lessons in which each individual STEM discipline served as a domain of enquiry with some level of integration from one or more instrumental domains.

11.4.2 Data Analysis

After collecting these lessons, we looked for trends in how educators from various content areas (science, technology, engineering, and mathematics) conceptualize and operationalize integrated STEM instruction. These analyses included not only the objectives for the lesson, but also the disciplinary knowledge and skills introduced as part of the instruction. This analysis took into account the *essential questions of curriculum* (Dillon, 2009), which was aimed at revealing the nature, elements, and practices of STEM curriculum developed and advocated by practitioners of various school subjects that compose STEM.

Using a clustering technique (Miles, Huberman, & Saldaña, 2014), we arranged each lesson into specific individual discipline categories based on the domain of enquiry. Once each lesson was categorized, we analyzed the articles to determine the disciplines acting as the instrumental domain, as well as if the knowledge and skills provided by the instrumental discipline were explicitly or implicitly connected by the teacher as part of the published lesson. The coding was performed by the first author, and then his findings were presented and discussed with the co-authors until agreement was achieved.

11.5 Key Findings

In examining the recent practitioner literature, it becomes apparent that operationalization of STEM education varies considerably depending on the curriculum developer's area of content expertise. A noticeable trend is disciplinary bias in the sense that STEM curricula tend to give primacy to the knowledge and skills of one specific discipline (domain of enquiry) within the STEM acronym while integrating parts from others (instrumental domains). To further elaborate on this point, a review of the literature in which one discipline is operationalized to a higher level above others is shown, beginning with science.

11.5.1 STEM in Science Education

Several examples of disciplinary dominance can be found within the science education literature. McHugh, Kelly, and Burghardt (2017) have middle school physical science students learn about thermal concepts by performing a directed inquiry lab using a heat transfer kit. Using Bybee's (2015) 5E approach, a STEM lesson was constructed to teach the concept of heat transfer, which was subsequently connected to mathematics instruction by having students calculate slope for temperature as a function of time. The authors argue that this lesson demonstrates a mathematics infusion, in which "mathematics content (is) taught in a science class where science is the major discipline and students apply mathematical skills and knowledge in science activities" (*ibid.*, p. 44). This praxis study acts as an example of a science highlighted approach that purports as STEM because the investigation is rooted in a science experiment, and there is some integration of mathematics principles. However, the lesson itself is somewhat devoid of engineering or technological principles. Furthermore, while the graphing of data and the calculation of mathematical slope are rooted in mathematics, their function as part of the lesson can also be considered part of a quantitative science investigation. The authors, of whom two of the three have a science disciplinary background, seem to conceptualize STEM as a way of integrating other skills for the benefit of the science discipline.

Similarly, Yoshikawa and Bartholomew (2018) propose using golf to promote technology and engineering principles. However, their proposed lesson revolves around studying the types of clubs and their purposes, and how to calculate the height of the ball using principles of classical physics. The only connection to engineering occurs when they propose that "golf could be used as a backdrop for learning to calculate the height of projectiles and practicing automation skills by creating a robotic to swing a 'club'" (*ibid.*, p. 35). While robotics certainly signifies a type of technology, it is used simply to replace a human from swinging a club. Furthermore, any engineering of a robot, including possible constraints, are missing from the proposed lesson. From that, it seems that science in the form of classical physics plays a central role in this lesson, and technology and engineering could be used as instrumental domains for integration to expand the science lesson rather than mere "decorative" elements in the problem space or scenario.

Trimble (2017) uses a project-based learning approach to teaching students about science communications. In this lesson, the students use free technology-based platforms to build science websites, such as Weebly and WordPress. In two different classes, the websites built involve persuasive arguments involving human effects on the environment as well as comparing and contrasting renewable energy companies. In addition, the students also built websites about a campus sustainability project, where they interviewed peers and uploaded their videos to the site. Trimble's integration of Internet website design highlights an integration of science as the domain of enquiry using technology resources as an instrumental domain, which is similar to the previously stated STEM lessons described.

Similarly, Smith, Roemmele, Miller, and Frisbee (2018) propose a problem-based learning (PBL) activity that seems to demonstrate the highlighting of one discipline within an integrated lesson. In this activity, the students were asked to design a roadway with a buffer zone that could absorb any chemical spills in case of an accident involving a truck with hazardous cargo. Students select materials to use in the buffer zone, and then test the porosity, infiltration, and levels of possible aquifer contamination using two-inch PVC pipes as test columns. After their tests, students then present their findings to the class. Overall, this lesson highlights a lot of science, and uses inquiry learning for students to select material for their proposed buffer zones. However, there is a slight integration of engineering in the design of the PVC pipe test column, but this use of engineering is only an instrument to test the students' designs of packed materials within the column. The actual selection of material involved guided inquiry experimentation of several possible sands and gravels, with a minimal constraint of water retention and slow motility through the selected material.

Schnittka (2017) provides an example of a more seamless integration of science and engineering, but one that lacks mathematics and technology. In a variety of lessons, Schnittka has her students design and build their own energy-making devices using different materials, such as magnets, steam production, a generator, or a wind turbine, among others. Further, students use multiple conversions to accurately calculate the amount of energy required to perform a function, allowing them to "see engineering as a way to solve problems, help people, and address a significant global environmental and health issue" (p. 42). This lesson constitutes an example of integration between science and engineering in which the boundaries begin to blur, and science principles and engineering constraints and specifications are taught concurrently. However, this is also an example of an integration that lacks mathematics, especially in dealing with the effects of friction and the loss of energy across multiple conversions.

Neilson and Campbell (2018) similarly propose a lesson that adds more mathematics and computational thinking to science. The authors present a five-day lesson using a wooden block on a ramp design. The wooden block has two sides–a smooth side and a side covered with sandpaper–being used by students to investigate the effect of the nature of surfaces on the normal force of the block on a ramp with a specific angle. The students build models and line graphs to demonstrate their experiments with their selected variables. Similar to the earlier presented work of McHugh et al. (2017) with their thermal kits, this science lesson has considerable incorporation of mathematical concepts. The mathematics component involves looking at the forces, an important part of a physics curriculum. It seems this lesson, while highlighting mathematics, still places scientific inquiry as the main domain of enquiry.

Olsen, Tofel-Grehl, and Ball (2018) provide another example of thermal energy transfer that focuses on science, but has loose connections to technology and engineering. In their lesson, the authors propose students build a lunch box that is better at insulating the contents inside. The authors provide some insulating materials, but "some teachers provide students with lunch boxes that are Mylar-lined or neoprene so students can explore the differences in insulation effects" (p. 49). In this manner, the students are not actively using the engineering design process to iteratively construct a prototype, but are merely using already known and constructed mass-produced lunch boxes. This limits the overlap between engineering, making the use of the lunchbox more of an example to demonstrate temperature change within an already constructed design. Furthermore, the temperature readings inside the lunchbox are taken using a microcontroller and LED panels that can be sewn on the box

Table 11.1 Practitioner articles highlighting science in integrated STEM lessons

References	Journal	Domain(s) of enquiry	Instrumental domain(s)	Explicit or implicit[a]
McHugh et al. (2017)	*Science Scope*	Science	Mathematics: calculating slope	E
Yoshikawa and Bartholomew (2018)	*Technology and Engineering Teacher*	Science	Technology and/or engineering: using a robot to swing the golf club	I
Trimble (2017)	*The Science Teacher*	Science	Technology: designing a website	E
Smith et al. (2018)	*The Science Teacher*	Science	Engineering: building a test column	I
Schnittka (2017)	*The Science Teacher*	Science and engineering	NA[b]	E
Neilson and Campbell (2018)	*The Science Teacher*	Science	Mathematics: plotting data and determining force	E
Olsen et al. (2018)	*Science Scope*	Science	Technology: using Arduino software to act as a thermometer	I

[a]Explicit or implicit instruction of knowledge or skills of the instrumental domain
[b]*NA* not applicable

to demonstrate when the temperature changes by increments of 5°*F*. Using Arduino software as a platform, "students can either be instructed to use the existing code or they can be provided with opportunities to modify it" (p. 50). While the use of Arduino software allows for a technology element, the fact that the code is already provided means that the students are simply provided with a recipe to set up a device that acts as a basic thermometer to take temperature readings. Because of these loose connections to engineering and technology, the lesson itself heavily favors the scientific principles that surround thermal energy transfer, one that is simply a modification of a science inquiry project in which students take internal temperature readings for a variety of provided lunchboxes.

In summation, the seven presented praxis lessons highlighting science in STEM education and how it exemplifies Gibbons' (1979) *integration by metamorphosis* framework is provided (see Table 11.1). As can be seen in Table 11.1, six of the seven practitioner lessons highlight science, which acts as the domain of enquiry. The instrumental domains shown are either explicitly taught as part of the lesson, or implicit as topics that can be added to enhance the lessons. There is considerable variation within this lesson set in terms of explicit or implicit instruction of the instrumental domain, where knowledge from one discipline is used to enhance the instruction of science. However, in each, the objectives, knowledge, and skills of one discipline is at the forefront of the lesson, and another discipline's knowledge or skill set is used as an enhancement. Only one lesson demonstrates a more equitable balance between science and engineering. As such, this demonstrates a lesson in which the lines begin to blur between the domain of enquiry and instrumental domain, and aligns more to a higher level of integration as defined by Vasquez's (2014) inclined plane of integration, as well as Moore's et al. (2014) description of content integration. Taken together, these lessons demonstrate a lack of clarity and coherency in not only the elements of an integrated STEM lesson, but also the degree of integration of one or more disciplines.

11.5.2 STEM in Technology Education

Integration of STEM highlighting one discipline also occurs within technology education practice. Technology itself has a somewhat ambiguous definition within the nebulous STEM umbrella, often shown as synonymous with engineering (Love, 2018). We chose to define technology as the use of new and emerging tools. This narrowed our focus to web-based applications, tablet usage, 3D-printing, robotics, and other digital technologies that are becoming more prominent in secondary lessons. In analyzing the literature categorized in this cluster, we found that the typical approach to integrated STEM education emphasizes technology design and use with the incorporation of mathematics and science principles. Cavanaugh and Trotter (2008) believe that "many educators see the study of technology as an opportunity to teach students how knowledge, tools, and skills in math and science can be applied to solve practical problems" (p. 18). Again, the view by these educators seems to demonstrate a highlighting of technology over the other disciplines, which are integrated only to enhance technological innovation.

Crippen and Archambault (2012) propose using online cloud-based data sets to help instruct what they describe as "STEM content." The authors give examples of building mash-ups, or visual representations of data, using Internet resources such as Yahoo! Pipe. These visuals are then used to promote argumentation skills for students, wherein the mash-ups can be used as evidence to cite claims and demonstrate justification. On the surface, this process seems to flexibly fit with the nebulous "STEM content" purported by the authors, but on closer inspection, much of the proposed lesson involves having students familiarize themselves with the technology. In short, although the scaffolded inquiry can benefit other disciplines, there still seems to be an overall main objective towards technological literacy.

Similarly, Hughes, Mona, Wilson, Seamans, McAninch, and Stout (2017) invite students to use 3D printers within the lesson. The authors promote several types of ways in which 3D printing can be used to facilitate a STEM lesson, including designing I beams and T beams for bridge design. However, the focus on the lesson is creating the file using 3D software, and not necessarily the engineering process itself. Consequently, this lesson seems to place technology at the center, where learning to utilize the instrument is the focus.

Robotics often plays a key role in enriching the technology discipline within integrated STEM lessons. Roman (2017) proposes a lesson having students research the costs of operating a greenhouse, and has them design how the layout of the greenhouse might change to have robots work to minimize cost. This lesson focuses more on the adding of technology to current ways of indoor growing and not necessarily manipulating scientific principles to help attain better crop yields. Consequently, this acts as an example of integrating technology and robotics for engagement in an unequal manner as compared to science.

Outside of robotics, mobile apps have also been advocated to enhance integrated STEM lessons. Bartholomew (2017) describes methods to incorporate the popular Pokémon Go app to integrated STEM classrooms. He proposes that the app can be used to teach mathematical principles, such as the calculation of distance between PokéStops, as well as science, in comparing and contrasting character traits to living flora and fauna. While these are certainly methods to integrate mathematics and science, the technology of the app is still at the focus and acts as the domain of enquiry, while science and mathematics supply an instrumental domain to enhance the app experience.

Technology can often play a significant role in engineering design challenges as well. Love and Ryan (2017) describe such an endeavor with the construction of miniature crab boats. The challenge, which was held in Maryland–a geographic locale known for its crab industry–uses regional interest to help engage the students to build replica crab boats at 1:12 scale, where one foot is proportioned down to one inch. Much of the challenge was designing the boat, and the authors recommended the use of 3D software to help in the design process, which links to technology. Depending upon how the lesson is implemented, there could be mathematics and scientific principles taught, as well as an iterative engineering design process. However, the challenge as a whole was to design and build the boat, which utilized more technology principles of design software and 3D printing of rudders and propellers. Therefore, while this lesson certainly has the capability of progressing further down the continuum of integration, its basic implementation focuses primarily on technology, with science, mathematics, and the engineering design process serving as ways in which further integration can be promoted depending upon how explicit these principles are implemented.

Technology need not always be cutting edge or computer-based when integrated, and the lesson provided by Kruse and Wilcox (2017) exemplifies this well. In their lesson, they describe a water purification process that is reminiscent of a science purification lab using filtration and activated charcoal. However, the lesson poses active questioning of students about the nature and philosophy of technology, and how processes such as filtration and items such as activated charcoal are technologies. This interesting take on a science lesson places technology in the foreground, allowing for technology and its nature to become the domain of enquiry while the science and engineering design of water purification became more of the vehicle for technological learning. Therefore, while this acts as an example of a more balanced integrative approach, it also clearly demonstrates technology as the lead discipline in an integrated STEM lesson.

Table 11.2 Practitioner articles highlighting technology in integrated STEM lessons

References	Journal	Domain(s) of enquiry	Instrumental domain(s)	Explicit or implicit[a]
Crippen and Archambault (2012)	*Computers in the Schools*	Technology	Science: evidence is scientific data	I
Hughes et al. (2017)	*Technology and Engineering Teacher*	Technology	Engineering: I and T beam construction	I
Roman (2017)	*Technology and Engineering Teacher*	Technology	Science: current layout of greenhouse based on science principles	I
Bartholomew (2017)	*Technology and Engineering Teacher*	Technology	Science and/or mathematics: comparing characters to local flora and fauna or measuring distance between PokéStops	E
Love and Ryan (2017)	*Technology & Engineering Teacher*	Technology	Science, mathematics, and/or engineering: principles of buoyancy, measurements for scale, iterative design process	I
Kruse and Wilcox (2017)	*Technology and Engineering Teacher*	Technology	Science: water purification and filtration	I

[a]Explicit or implicit instruction of knowledge or skills of the instrumental domain

In summation, the six practitioner lessons presented all highlight technology as the domain of enquiry according to Gibbons' (1979) theoretical framework (see Table 11.2). In each of these lessons, the main objectives of technology are presented within the lesson, whether it be using a 3D printer or an Internet resource. Similar to the findings presented earlier for science, there is again variation in the implementation of the other disciplines. The crab boat challenge presented by Love and Ryan (2017) demonstrates a more equitable balance of integration, however the explicit implementation of science, mathematics, or engineering elements would greater enhance the level of integration according to Vasquez's (2014) inclined plane of integration. These lessons demonstrate more of the context integration proposed by Moore et al. (2014), in which the goals of one discipline–in this case, technology–is imbalanced as part of the integrated lesson. Furthermore, the level of integration and the disciplines involved greatly differ, which exemplifies the lack of coherence and clarity that is found in the practitioner literature for integrated STEM lessons.

11.5.3 STEM in Mathematics Education

Mathematics has also seen itself highlighted as part of an integrated STEM approach in the recent literature. As a former head of the National Council of Teachers in Mathematics, Shaughnessy (2013) argues that STEM approaches should involve "significant mathematics in the problem" (p. 324). An example of this is provided by Smith, Seshaiyer, Peixoto, Suh, Bagshaw, and Collins (2013) who describe a STEM lesson in which students calculate slope as rate of change using the lengths and heights of staircases. The lesson itself affords an opportunity for students to actively measure real-world staircases and calculate the slope as an application of a tangible entity. The authors argue that technology has been introduced in the form of computer programming to form graphs. They also state that engineering practices are in place because they involve the application of stairs. However, the lesson itself does not involve the actual engineering design and building of stairs. Consequently, this also acts as an example of a STEM problem that highlights one discipline over others.

Another example of a STEM lesson that highlights mathematics is proposed by Magiera (2013) and involves the use of model eliciting activities (MEAs). These activities begin with a reading passage that situates the real-world context of a mathematics problem, and students use critical thinking skills to design a procedure to find a mathematical solution to the posed problem. While the reading passage allows for the integration of the other STEM disciplines, the lesson at its core focuses on performing mathematical calculations for a desired conclusion. More specifically, students use area to count the number of metal crystals in three images of aluminum, and connect the amount of crystals in the contained area to strength. While the lesson does provide the students with a way to design their own procedure, the author states that "students experience mathematical and scientific inquiry because MEAs give students problem-solving experiences that are similar to those of scientists and engineers" (p. 354). While the experience of problem-solving is similar, the lack of constraints and specifications necessary in engineering as well as the connection to scientific principles of tensile strength limited this lesson to heavily favor mathematics instruction over other integrated disciplines within the STEM acronym.

Mathematics was also found to be integrated with technology as part of a STEM lesson. Fujiwara (2018) describes a calculus-based lesson in which the volume of a solid revolving around an axis can be found mathematically, and is then compared to a 3D printed shape of the same equation. While the use of a 3D printing source serves to integrate technology into the lesson, the equation itself is provided to the student. Therefore, the mathematical principle of volume calculation serves as the domain of enquiry, while technology is integrated to produce a tangible representation of the mathematical concept. In short, the technology is simply an instrument to help further the mathematical objectives of this lesson, which are foregrounded above all others in this form of integration.

Technology is also used as an instrumental domain in the lesson proposed by Pope (2018). In this lesson, students throw a beanbag into a bowl across the room, and videotape the arc of the bag as it approaches the target. This video is then uploaded into Vernier's *LoggerPro*® software, and then used to determine the quadratic equation that matches the movement of the beanbag. This lesson highlights the integration of technology in a mathematics-based lesson, and can also be extended to the scientific principles involving projectile motion in physics. Consequently, it provides another example of an integrated lesson that shows an imbalance between integrated disciplines.

Similarly, Eddy, Pratt, and Green (2018) present a lesson that also integrates technology to serve mathematical learning. Their lesson involves the integration of Google® Maps and Desmos to help aid in teaching linear inequalities. In it, Google® Maps is used to map out points of interests in the local town, and students use linear equations to help identify streets based upon slope and where it crosses the designated *y* and *x* axes. The students then use inequalities to shade areas that are less than or greater than two shown roads, which have linear equations determined for them by the students. The students then use this to find where a particular point of interest might lie in the overlapping areas. This lesson provides a real world context to the instructed mathematics principles, and utilizes technology as an instrumental domain to help facilitate engagement.

Mathematics learning in real world context is also a focus in the lesson devised by Alhammouri, Foley, and Dael (2018). In their lesson, they use the context of stocking a lake with trout to connect with statistics at the secondary level, namely the linear difference equations and the concept of reaching equilibrium. In the lesson, the students use the modeling cycle framework (Bliss, Fowler, and Galluzzo, 2014) to justify how a trout population might change over a period of time in a lake that is annually stocked. The students use the cycle, which incorporates asking questions and building models, skills consistent with the scientific and engineering practices of the Next Generation Science Standards (NGSS Lead States, 2013). A group discussion is then facilitated to help answer the questions, from which the students built linear difference equations to help answer the models that they built to approach the mathematics problem. For technology integration, the students use TI-*n*spire graphing calculators. The integration of technology as well as a modeling process embraced by science and engineering demonstrates this lesson as one that has elements of integration, albeit one in which the goals of the mathematics discipline are at the center of the lesson itself.

In summation, the six practitioner lessons presented demonstrate integrated lessons in which there is imbalance that favors mathematics as part of the integration. In all of these lessons, the domain of enquiry is mathematics, but there are differences in the types of disciplines being integrated (see Table 11.3). Furthermore, in each lesson, there is only implicit use of the principles and skills of the integrated instrumental disciplines. Therefore, each of these demonstrates lower levels of integration, such as the context integration posed by Moore et al. (2014). They also demonstrate the lack of coherence in integrated STEM lessons, as the types of

Table 11.3 Practitioner articles highlighting mathematics in integrated STEM lessons

References	Journal	Domain(s) of enquiry	Instrumental domain(s)	Explicit or implicit[a]
Smith et al. (2013)	*Mathematics Teaching in the Middle School*	Mathematics	Technology: making computer graphs	I
Magiera (2013)	*Mathematics Teaching in the Middle School*	Mathematics	Science & engineering: inquiry and iterative design	I
Fujiwara (2018)	*Technology & Engineering Teacher*	Mathematics	Technology: building a 3D model of an equation	I
Pope (2018)	*Mathematics Teacher*	Mathematics	Technology: use of LoggerPro®	I
Eddy et al. (2018)	*Mathematics Teacher*	Mathematics	Technology: use of GoogleMaps®	I
Alhammouri et al. (2018)	*Mathematics Teacher*	Mathematics	Technology: use of TI-*n*spire graphing calculators	I

[a]Explicit or implicit instruction of knowledge or skills of the instrumental domain

disciplines integrated vary as well as the degree of the knowledge and skills utilized to enhance the lesson.

11.5.4 STEM in Engineering Education

Finally, engineering is highlighted over other disciplines in many integrated STEM lessons. The discipline of engineering itself involves the design and building of equipment to solve real-world problems based upon constraints and specifications. Because of this, there does appear in the literature ample cases where the design process in engineering is utilized in conjunction with one of the other disciplines. An example of this is English and Mousoulides' (2015) middle-school lesson in which students are asked to design a bridge to replace the one that fell in Minneapolis in 2007. The students have to research different types of bridges, and utilize mathematics principles to ensure the bridge can support the proper weight and have the appropriate area to fit various bridge designs. Similarly, Enderson and Grant (2013) use the design process in constructing paper tables. The students build models, hold discussions, and use an iterative process with redesign to build paper tables that can hold a given amount of weight. While the students are able to incorporate mathematical principles of strength of shape, the focus of the inquiry is on the design process itself. Taken together, both of these are examples in which engineering is a focus, and math is utilized to meet the goals of the engineering discipline itself.

Another example in which engineering is at the center of the lesson is Welling and Wright's (2018) lesson involving the construction of paper rockets. In this lesson, students practice the engineering design process (EDP) by building rockets out of paper, which are inexpensive and allow for "students to create and evaluate five designs in one class period" (p. 19). This helps highlight the design process, allowing for multiple iterations of design. However, the design process itself only loosely connects to science and mathematical principles. In the second phase of the project, the authors describe student-led focus groups that have the students use experimentation to identify one of ten different variables, ranging from nose cone length to number and size of fins. While this loosely connects to scientists' utilization of dependent and independent variables, the lack of specific instruction on aerodynamics and other scientific connections means that most of the design enhancements were made by trial-and-error. Therefore, this lesson yet again demonstrates one that places engineering at the center, with implicit scientific principles not readily taught by the instructors.

Similarly, Hemming (2018) proposes a lesson involving the construction of drawbridges. He provides constraints for the amount of mass required for the bridge to hold as well as the scaled size. In the exploration phase of Bybee's 5E approach (2015), Hemming has the students use a force lab provided by an online resource to build simulated bridges. Again, this exploration seems to utilize a trial-and-error approach to bridge building that has a lack of connection to scientific principles of strength of material as well as mathematical connection to geometry and shape strength. As such, this lesson demonstrates integration in which the engineering design process is placed at the center of a STEM lesson.

Hughes, Mona, Stout, and McAninch (2015) supply another example in which the engineering design process is at the forefront of an integrative STEM lesson. The authors present a lesson involving the collection of solar energy to create a water heater that has reduced usage of fossil fuels. However, most of the lesson surrounds designing the water heater itself. There is some integration of science principles, namely the energy absorption of black objects. In a later iteration of the project, the authors describe building more "close-looped systems, (where) pressure gauges were added for safety and students were introduced to Charles' and Boyle's laws" (p. 29). This demonstrates the incidental learning of a science principle to help facilitate the engineering design. This incidental learning also occurs in mathematics, in which the authors have the students use pre-made equations to help find the optimal angle to point their water heaters at the sun based upon latitude. However, students simply use the equation and Microsoft Excel® to plot the data without truly investigating the underlying mathematical relationships that comprise the formula, thus placing engineering yet again as the focus of the integrative STEM lesson.

The examples above primarily demonstrate when STEM lessons are in some ways enhanced with science, but there is a lack of equality in the integration process. The next lesson is one in which science and engineering are more equitably integrated, but there lacks a connection to mathematics as well as technology. Wheeler, Whitworth, and Gonczi (2014) introduced a lesson in which students build a voltaic cell that can power a small motor and an attached fan. Students are given the constraint to build the voltaic cell using only provided materials that produces enough electrical energy to power a small fan. They then have the opportunity to research the science principles surrounding electrochemical cells and cell potential, while also delving into air resistance and mass in the design of the fan. Throughout the lesson, students use the research to iteratively design their cell-motor apparatus to meet the constraints provided. This demonstrates more of an equitable balance between science and engineering in a STEM lesson.

Similarly, Gerber, Halsted, Hershberger, Riddle, Foster, and Hill (2018) provide a more recent example of another STEM lesson that more equitably balances science and engineering at the middle school level. In their lesson, the authors propose the construction of a prosthetic hand. The lesson begins by students using the 5E approach to explore the scientific principles surrounding the anatomy of the hand, including muscles and tendons. The students then utilize this scientific knowledge to build models made from paper and string to demonstrate finger movement and dexterity. From this, constraints for durability, flexibility, weight load, and ability to resist water were proposed, and students constructed example hands using household materials provided by the instructors that fit these basic constraints. Students even have the ability to test and modify their designs depending upon time constraints, lending itself to the more iterative nature of engineering design. While the lesson itself lacks a firm connection to mathematics and technology, it shows on the whole a more equitable integrative balance between science and engineering, where the principles of anatomy were made explicit and the constraints of the engineering design were fostered by the instructors upon based upon these principles.

In summation, the seven practitioner lessons presented demonstrate situations in which the principles of engineering–such as the iterative design process and real-world contextual problems–acts as the main inquiry. The other STEM disciplines act in ways to enhance the engineering process, acting as instrumental domains (see Table 11.4). As can be seen, there is again a lack of coherence in the type of discipline integrated as well as the degree in which the knowledge and skills of those are applied within the lesson. Finally, there is also variation in the extent of explicit instruction of the enhancing disciplines, which affects the level of integration as well. Taken together, this demonstrates a lack of uniformity, which demonstrates the lack of clarity in conceptualizing integrated STEM education lessons by the designers and practitioners of these lessons.

11.6 Discussion

In looking across the above pedagogical approaches, the diversity and variation in what constitutes STEM is promptly noticeable. However, it would be unfair to disparage the educational value of any of these lessons. All of them act as wonderful examples of effective current teaching strategies, and have been included because of

Table 11.4 Practitioner articles highlighting engineering in integrated STEM lessons

References	Journal	Domain(s) of enquiry	Instrumental domain(s)	Explicit or implicit[a]
English and Mousoulides (2015)	*Mathematics Teaching in the Middle School*	Engineering	Mathematics	E
Enderson and Grant (2013)	*Mathematics Teaching in the Middle School*	Engineering	Mathematics: strength of shape	E
Welling and Wright (2018)	*Technology and Engineering Teacher*	Engineering	Science: variables affecting aerodynamics	I
Hemming (2018)	*Technology and Engineering Teacher*	Engineering	Technology: online simulator	I
Hughes et al. (2015)	*Technology and Engineering Teacher*	Engineering	Science & mathematics: energy absorption of black objects/gas laws and optimal angle of the sun	E
Wheeler et al. (2014)	*The Science Teacher*	Engineering	Science: electrochemical principles	E
Gerber et al. (2018)	*Science Scope*	Engineering	Science: anatomy of a hand	E

[a]Explicit or implicit instruction of knowledge or skills of the instrumental domain

their utility and novelty as well as their example of STEM integration. Nevertheless, it is important to note the variation in what teachers call "STEM" and the inequality between the integrated disciplines that occurs within each individual lesson.

Integrated curriculum recently created by practitioners not only varies considerably in content and process, but it is commonly characterized by disciplinary bias and an imbalanced focus. Considerable ambiguity also exists with regard to what constitutes various levels of integration as well as what the boundaries of each discipline are. Further, curriculum developers' conceptions of what it means for disciplinary borders to dissolve remain quite nebulous. Consequently, we believe that a more comprehensive review of praxis and conceptualization of integrated STEM education in practitioner journals would be warranted and most beneficial to the field. Such an endeavor can help clarify what exactly constitutes the meta-discipline of STEM and advance our understanding of how to effectively integrate curricula.

Gibbons' (1979) *integration by metamorphosis* speaks of integration in epistemological sets of two through the domain of enquiry and the instrumental domain. Upon close inspection, a majority of the above cited praxis examples fit this mode of integration, in which one discipline is highlighted over the others. Moore et al.

(2014) calls this type of integration context integration, and it represents a lower level according to Vasquez's (2014) inclined plane of integrated STEM lessons. What appears in looking at the corpus of our presented articles, there is a pattern of variation between lessons in both the types of disciplines used to enhance the main inquiry, the degree of integration, and the explicit or implicit instruction of knowledge and elements of those instrumental domains. Overall this demonstrates a lack of coherence in what constitutes an integrated STEM lesson, as well as the need for greater clarity in the elements and design of an exemplar integrated STEM lesson.

The observations made from these lessons pose the following questions: Why are the STEM lessons so varied? Why does it seem that one discipline acts as the main inquiry? It is our thesis that these questions arise as a direct result of the ambiguous and nebulous manner in which STEM education is conceptualized. For the remainder of this section, we discuss where this ambiguity possibly comes from, as well as some current empirical strategies posed in the literature to help counteract the varying conceptualizations that seem to permeate the integrated STEM education literature.

11.6.1 Competing Agendas in STEM Education

Akerson, Burgess, Gerber, Guo, Khan, and Newman (2018) argue that the very nature of science, technology, engineering, and mathematics are vastly different, and "STEM itself is a socially constructed label that is in response to economic and global pressure. It has always existed but was simply the individual disciplines that compose it, influencing and building on one another" (p. 5). Herein, Akerson et al. cite the significant political influence that surrounds STEM and STEM education as a whole.

Consistent with this argument, most empirical articles as well as position papers cite early in their introductions the urgency of STEM and STEM education, not only for producing twenty-first century workers for STEM fields (Presidents' Council of Advisors on Science and Technology, 2010), but also in the United States falling behind the rest of the world in producing students ready to compete within fields (National Academies, 2006). The politically charged nature of STEM therefore adds a certain complexity and variation to the amount of stakeholders within the field of STEM education. Not only do parents, students, administrators, and teachers have a stake in STEM education, but politicians as well. This ultimately leads to differing opinions about what constitutes a successful STEM education program. Politicians may view STEM education as a way to advance a workforce and invest in a strong economic future, while individual teachers might view STEM as a means to help incorporate real-world context into their own curricula. The consequence of these competing goals and views leads to different ways of conceptualizing STEM education, and therefore different methods of integration and incorporation of the individual disciplines. In short, the competing agendas of all stakeholders' cloud how STEM is viewed, allowing for differing interpretations to co-exist under the

STEM education umbrella and leading to an ambiguity into what constitutes effective STEM education.

11.6.2 Teacher Biased Expertise

This chapter has sought to illuminate the ambiguity of integrated STEM education praxes that currently exist in the field. The examples of lessons demonstrate the variations in focus of STEM. While the political influence does play a role in the nebulous nature of STEM education, the experience and background of the instructor also has an impact in the lesson as well. Many of the lessons in which there was a more balanced integrated approach demonstrate explicit instruction of instrumental domains as a means of accomplishing the main inquiry of the lesson. The empirical literature supports the need for explicit instruction as well. Berland and Steingut (2016) found that students inconsistently used scientific and mathematical principles when performing engineering design challenges with an integrated STEM education environment, instead relying upon a trial-and-error approach. They propose that instruction of the requisite mathematics and science necessary to perform the function must be explicitly taught by the instructors, and "to tie the engineering design project and target content together so tightly that it is impossible to attend to one without the other" (p. 2756). This strategy is consistent with the upper levels of transdisciplinarity in the inclined plane of STEM integration proposed by Vasquez (2014) earlier.

Explicit instruction is not without its own challenges, especially in an integrated STEM lesson. Because STEM itself is comprised of four disciplines, teachers themselves must be well versed not only in their own discipline, but also in the others. However, most teachers do not have an equitable amount of education and professional experience in all four facets of STEM. For example, one of us (Sgro) is currently a high school science teacher. His background is in chemistry, and the courses that he teaches are about chemistry. When taking an integrated approach, he would add technology, engineering and mathematics principles to his lessons, but the end result of those lessons would still center upon the science of chemistry itself. This could ultimately lead to an unbalanced approach to his integrated STEM lessons, in which one discipline acted as the main domain of enquiry. In what we see in the above cited literature, this occurs for a multitude of other teachers as well.

This inequality when it comes to integration seems to occur for two main reasons: factors beyond the teacher's control and the teacher's own educational background and teaching experiences. First, in any public primary or secondary education, there are a host of factors involved in instruction beyond the teacher's control. These include often a set curriculum provided by a local or state agency, high-stakes standardized exams that occur at the conclusion of the school year, and the time constraints that exist within a standard instructional school day. In any lesson design, these three factors ultimately have an impact in designing the overall student learning objectives of the integrated lesson as well as the pedagogical methods used. Therefore, within the time constraints, the teacher will often have to make judgements about what parts of lessons that he or she wishes to highlight as well as what parts might be eschewed to help streamline the lesson and not obscure the selected educational outcomes. Ultimately this can lead to an imbalance in integrated approach, especially for STEM lessons. While an imbalance in and of itself is not necessarily detrimental to a lesson, it does cloud the essential elements of what constitutes a successful integration, especially in STEM, in which four different discipline's epistemological and ontological views are mixed to create a lesson that transcends their individual boundaries. In addition, when one discipline is placed with greater importance, it implicitly demonstrates that that discipline's knowledge and process may be more valued than others. This could lead to a replicating of bias towards one discipline by students as well. Therefore, while it may be difficult to balance all disciplines in every lesson, having a greater call for greater equity of integration in all lessons would not only benefit the students in their valuations of the parts that comprise STEM, but also create opportunities for shared conceptualizations of STEM education by its practitioners.

Recently, these issues that affect teachers have been addressed in the empirical literature. Smith, Parker, McKinney, and Grigg (2018) found that when elementary teachers in urban public schools were given a pre-packaged integrated STEM curriculum, there was great variation in the implementation process. The variation in implementation occurred because of not only time constraints, but also the teacher's perceptions in their students' abilities as well as the teacher's understanding of the main themes of the lesson itself. Because of this, students in different classes had different parts of their lessons highlighted, creating a scenario for imbalance within the integrated lesson to occur.

A teacher's ability to understand the lesson itself is part of the second reason in which there is such imbalance in integrated STEM lessons. A teacher enters the class with certain pre-conceived notions of their discipline. As a professional, the teacher has been educated in a community of practice (Wenger, 1998). For their specific discipline, these teachers are versed in various philosophical assumptions as well as epistemological and ontological outlooks when it comes to not only their specific discipline, but education and the educational process as a whole. These foundations as well as subsequent experiences set the tone for how the teacher implements his or her lessons. Therefore, in approaching an integrated STEM lesson, a teacher versed in science is going to focus on science, implementing the other disciplines as needed. Moreover, the more shallow the foundation and understanding a teacher might have in one of the integrated disciplines, the less likely explicit integration might take place. A teacher can only explicitly call attention to real-world connections between integrated STEM disciplines if he or she is *aware* of those connections to begin with. Furthermore, the more fluent the instructor might be in other disciplines, the more likely he or she will draw upon that information when designing a lesson or be able to shift when a teachable moment surfaces within the lesson itself.

Teacher background and its impact on integrated STEM instruction is prevalent in the literature as well. Akerson et al. (2018) allude to the fact that even their authors have varying backgrounds and conceptualize STEM differently because of their varied educational foundations and ways of knowing. Dare, Ellis, and Roehrig (2018) found that in a cross-case phenomenological study that integration across STEM disciplines occurred at varying percentages and that "the degree of implementation may be related to teachers' awareness of how to make explicit and meaningful connections between the disciplines" (p. 17). Additionally, Wang, Moore, Roehrig, and Park (2011) found in a multiple case study that the teacher's background played a large role in their enactment of the integrated curriculum. In their cases, the science teacher conceptualized integrated STEM as more problem-solving based, while the mathematics teacher perceived integrated STEM as giving real-world context to mathematics. The engineering teacher perceived STEM as a way of using science and mathematics principles to approach an engineering design challenge. Consequently, their differing perceptions as well as backgrounds allowed for differing levels of integration, different pedagogical approaches, and ultimately, a different focus for their lessons. Taken together, these empirical studies demonstrate an unequal balance that occurs within an integrated STEM lesson, findings that corroborate the praxis articles that we have surveyed from the practitioner literature as well.

There are possible solutions to the impact of a teacher's own bias and educational background. Integrated curricula could benefit by the implementation of team-teaching or co-teaching. In this way, having teachers with different backgrounds coming together can help support greater levels of integration. An example of this would be a STEM lesson taught in conjunction by a science teacher, and engineering teacher, and a mathematics teacher. This could be done within the confines of one course or three courses that have a shared lesson. In either scenario, the objectives of each of those three disciplines can be highlighted, as well as the incorporation of new and emerging technologies.

11.6.3 Toward Greater Clarity in Integrated STEM Lessons

If the goal of an integrated approach is to produce students with STEM skills and knowledge, then having an understanding of what constitutes those specific skills is important. Kennedy and Odell (2014) describe this as STEM becoming a "meta-discipline," defined as "an integrated effort that removes the traditional barriers between these subjects, and instead focuses on innovation and the applied process of designing solution to complex contextual problems using current tools and technologies" (p. 246). While their view of the STEM meta-discipline focuses on integration of technology and engineering into current coursework, the idea of having a designated set of philosophical assumptions, epistemology, ontology, and semiotic practices for "STEM education" could benefit by adding clarity to the idea of an integrated STEM education approach. In short, stakeholders coming together to identify the salient pieces of an integrated STEM program could not only help clarify what "STEM" is, but allow for a greater uniformity in the types of praxis that can be designated as "STEM."

What are the main features of integrated STEM education? The answer to this often depends upon the stakeholder, but the example praxis lessons provided earlier demonstrate a pattern that highlights salient features of what currently constitutes an integrated STEM lesson. First, the lesson is student-centered, often involving problem-based learning (Hmelo-Silver, 2004). Second, the problem seems to exist within a real-world context. Put simply, the problem at the center of the inquiry is one that exists as a complex problem that currently exists in the world, whether it is a local issue or one of global significance. Beyond these features, integrated STEM lessons tend to differ. It is these differences that would benefit from clarity and coherence in not only defining integrated STEM education, but also in assembling a set of skills, knowledge, and practices that encapsulate a unit or lesson.

Recently, there have been identifiable strategies in the educational literature to help achieve this call for clarity and coherency in STEM education. One example is the conceptual framework postulated by Kelley and Knowles (2016). Their idea of a conceptual framework likens situated STEM learning as a set of pulleys that represent science, engineering, technology, and mathematics. In their visual representation, science and engineering represent larger pulleys than technology and mathematics, which somewhat signifies a viewpoint that puts more emphasis on science and engineering. In approaching this framework from Gibbons' theory of integration by metamorphosis, it would appear that Kelley & Knowles see science and engineering as the domains of enquiry, and technology and mathematics can be used as instrumental domains to help enhance the iterative design process and its guided utilization of scientific principles. Moving through and connecting these pulleys is the idea of a set community of practice that interweaves through each discipline. They state that "foundational to this theory is the concept that understanding how knowledge and skills can be applied is as important as learning the knowledge and skills itself" (p. 4). While there is still some ambiguity and inequality in what constitutes STEM education in their conceptual framework, overall, it provides a unique starting point to start dialogue between stakeholders about common features of an integrated STEM education approach.

Similar to the idea of a conceptual framework, Chalmers, Carter, Cooper, and Nason (2017) have recently proposed the implementation of a set of "big ideas" to help clarify an integrated STEM approach. Building off previous research on "big ideas" in the individual disciplines (Askew, 2013; Harlen, 2010), they propose a set of big ideas that encompass the overlap of the four disciplines to be used to help design and implement a more coherent form of integrated STEM education. Their proposed big ideas would surround three main functions: within-disciplinary big ideas that can be applied to the other disciplines, cross-discipline ideas that are in two or more areas of STEM, and encompassing big ideas that use all of the disciplines to approach problem solving. The idea is that using these "big ideas" can help teachers design units, especially for teachers that may have more background in one specific discipline. In short, these proposed ideas can help clarify and add a coherency to the types of STEM praxis that can occur in primary and secondary schools, leading to less of an imbalance between the individual disciplines and affording an opportunity for greater connections between praxis lessons that purport to be integrated STEM endeavors.

11.7 Concluding Thoughts

We would like to conclude by pointing out that the problem of curricular imbalance is far from novel or unique to integrated STEM teaching efforts. Throughout the history of American education, teachers have sought to strike balance between epistemological product (knowledge) and process (method), between teacher- and student-centeredness, between summative and formative assessment, to name a few. Some have characterized this state of affairs as a metaphorical "pendulum" perennially swinging back and forth without ever reaching stationary equilibrium. Curriculum theorists like Ellis (2004) point out that "balance is seldom achieved in any curriculum, no matter how it is structured" (p. 61) and "the search for curricular balance is never easy and it remains one of the great challenges to any teacher" (p. 106). Solving this conundrum is imperative if STEM educators are to succeed in their efforts to provide students with truly transformative learning experiences that are less siloed and less constrained by disciplinary boundaries.

References

Akerson, V., Burgess, A., Gerber, A., Guo, M., Khan, T. A., & Newman, S. (2018). Disentangling the meaning of STEM: Implications for science education and science teacher education. *Journal of Science Teacher Education, 29*(1), 1–8. https://doi.org/10.1080/1046560X.2018.1435063

Alhammouri, A. M., Foley, G. D., & Dael, K. (2018). Tracking trout: Engaging students in modeling. *Mathematics Teacher, 111*(6), 416–423.

Askew, M. (2013). Big ideas in primary mathematics: Issues and directions. *Perspectives in Education, 31*(3), 5–18.

Bartholomew, S. R. (2017). Using Pokémon go to teach integrative STEM. *Technology and Engineering Teacher, 76*(5), 24–27.

Becker, K., & Park, K. (2011). Effects of integrative approaches among science, technology, engineering, and mathematics (STEM) subjects on students' learning: A preliminary meta-analysis. *Journal of STEM Education, 12*(5 & 6), 23–37.

Berland, L. K., & Steingut, R. (2016). Explaining variation in student efforts towards using math and science knowledge in engineering contexts. *International Journal of Science Education, 38*(18), 2742–2761. https://doi.org/10.1080/09500693.2016.1260179

Bliss, K. M., Fowler, K. R., & Galluzzo, B. R. (2014). *Math modeling: Getting started and getting solutions*. Philadelphia, PA: Society for Industrial and Applied Mathematics.

Breiner, J., Harkness, M., Johnson, C. C., & Koehler, C. (2012). What is STEM? A discussion about conceptions of STEM in education and partnerships. *School Science and Mathematics, 112*(1), 3–11.

Bybee, R. W. (2015). *The BSCS 5E instructional model: Creating teachable moments*. Arlington, VA: NSTA Press.

Cavanaugh, S., & Trotter, A. (2008). Where's the T in STEM? *Education Week, 27*(30), 17–19.
Chalmers, C., Carter, M., Cooper, T., & Nason, R. (2017). Implementing "big ideas" to advance the teaching and learning of science, technology, engineering, and mathematics (STEM). *International Journal of Science and Math Education, 15*(Suppl 1), S25–S43. https://doi.org/10.1007/s10763-017-9799-1
Crippen, K. J., & Archambault, L. (2012). Scaffolded inquiry-based instruction with technology: A signature pedagogy for STEM education. *Computers in the Schools, 29*, 157–173.
Dare, E. A., Ellis, J. A., & Roehrig, G. H. (2018). Understanding science teachers' implementations of integrated STEM curricular units through a phenomenological multiple case study. *International Journal of STEM Education, 5*(4), 1–19. https://doi.org/10.1186/s40594-018-0101-z
Dillon, J. T. (2009). The questions of curriculum. *Journal of Curriculum Studies, 41*(3), 343–359.
Eddy, C. M., Pratt, S. S., & Green, C. (2018). Interactive maps for systems of linear-inequalities. *Mathematics Teacher, 111*(6), 425–430.
Egan, K. (1978). What is curriculum? *Curriculum Inquiry, 8*(1), 65–72.
Ellis, A. K. (2004). *Exemplars of curriculum theory*. Larchmont, NY: Eye on Education.
Enderson, M. C., & Grant, M. R. (2013). Emerging engineers design a paper table. *Mathematics Teaching in Middle School, 18*(6), 362–369.
English, L. D. (2016). STEM education k-12: Perspectives on integration. *International Journal of STEM Education, 3*(3), 1–8.
English, L. D., & Mousoulides, N. G. (2015). Bridging STEM in a real-world problem. *Mathematics Teaching in the Middle School, 20*(9), 532–539.
Fujiwara, Y. (2018). STEM integration: Solids, CADs, and 3D printers. *Technology and Engineering Teacher, 77*(8), 5–9.
Gerber, A., Halsted, A., Hershberger, M., Riddle, L., Foster, K., & Hill, A. (2018). Engineering prosthetic hands: An activity for integrating STEM in the science classroom. *Science Scope, 42*(3), 66–73.
Gibbons, J. A. (1979). Curriculum Integration. *Curriculum Inquiry, 9*(4), 321–332.
Harlen, W. (Ed.). (2010). *Principles and big ideas of science education*. Hatfield, UK: Association of Science Teachers.
Hemming, J. (2018). Drawbridge by design: Civil engineering for middle school. *Technology and Engineering Teacher, 77*(7), 40–44.
Hmelo-Silver, C. E. (2004). Problem-based learning: What and how do students learn? *Educational Psychology Review, 16*(3), 235–266.
Honey, M., Pearson, G., & Schweingruber, H. (2014). *STEM integration in K-12 education: Status, prospects, and an agenda for research*. Washington, DC: National Academy of Sciences.
Hudson, P. (2015). Science, technology, engineering, and Maths (STEM). In R. Gunstone (Ed.), *Encyclopedia of science education* (pp. 940–941). Dordrecht, The Netherlands: Springer.
Hughes, B., Mona, L., Stout, H., & McAninch, S. (2015). An integrative STEM approach to teaching solar energy collection. *Technology and Engineering Teacher, 75*(1), 26–31.
Hughes, B., Mona, L., Wilson, G., Seamans, J., McAninch, S., & Stout, H. (2017). Every day a new 3D printing material. *Technology and Engineering Teacher, 76*(5), 8–13.
Kelley, T. R., & Knowles, J. G. (2016). A conceptual framework for integrated STEM education. *International Journal of STEM Education, 3*(11), 1–11. https://doi.org/10.1186/s40594-016-0046-z
Kennedy, T. J. & Odell, M. R. L. (2014). Engaging students in STEM education. *Science Education International, 25*(3), 246–258.
Kruse, J., & Wilcox, J. (2017). Using a water purification activity to teach the philosophy and nature of technology. *Technology and Engineering Teacher, 76*(8), 13–19.
Kysilka, M. L. (1998). Understanding integrated curriculum. *The Curriculum Journal, 9*(2), 197–209.
Love, T. S. (2018). The T&E of STEM: A collaborative effort. *The Science Teacher, 86*(3), 8–10.

Love, T. S., & Ryan, L. (2017). The crab boat engineering design challenge. *Technology and Engineering Teacher, 76*(7), 8–14.
Magiera, M. T. (2013). Model eliciting activities: A home run. *Mathematics Teaching in the Middle School, 18*(6), 348–355.
McHugh, L., Kelly, A. M., & Burghardt, M. D. (2017). Teaching thermal energy concepts in a middle school mathematics-infused science curriculum. *Science Scope, 41*(1), 43–50.
Miles, M. B., Huberman, A. M., & Saldaña, J. (2014). *Qualitative data analysis: A methods sourcebook* (3rd ed.). Thousand Oaks, CA: Sage.
Moore, T. J., Stohlmann, M. S., Wang, H.-H., Tank, K. M., Glancy, A. W., & Roehrig, G. H. (2014). Implementation and integration of engineering in K-12 STEM integration. In S. Purzer, J. Strobel, & M. Cardella (Eds.), *Engineering in precollege settings: Research into practice* (pp. 35–60). West Lafayette, IN: Purdue Press.
National Academies. (2006). *Rising above the gathering storm: Energizing and employing America for a brighter economic future*. Washington, DC: National Academies Press.
Neilson, D., & Campbell, T. (2018). Adding math to science. *The Science Teacher, 86*(3), 26–32.
NGSS Lead States. (2013). *Next generation science standards: For states, by states*. Washington, DC: The National Academies Press.
Olsen, E., Tofel-Grehl, C., & Ball, D. (2018). The temperature-sensing lunchbox. *Science Scope, 42*(1), 47–52.
Pinnar, W. (2011). *The character of curriculum studies: Bildung, currere, and the recurring question of the subject*. New York, NY: Palgrave Macmillan.
Pope, D. (2018). Exploring quadratic functions using logger pro. *Mathematics Teacher, 111*(6), 454–460.
Presidents' Council of Advisors on Science and Technology. (2010). *Prepare and inspire: K-12 education in science, technology, engineering, and math (STEM) education for America's future*. Retrieved from https://nsf.gov/attachments/117803/public/2a%2D%2DPrepare_and_Inspire%2D%2DPCAST.pdf
Radloff, J., & Guzey, S. (2016). Investigating Preservice STEM teacher conceptions of STEM education. *Journal of Science, Education, and Technology, 25*, 759–774.
Roman, H. T. (2017). The greenhouse robot challenge. *Technology and Engineering Teacher, 77*(3), 42–43.
Sanders, M. (2009). STEM, STEM education, STEMmania. *The Technology Teacher, 68*(4), 20–26.
Schnittka, C. (2017). Gravity can do what? *The Science Teacher, 84*(8), 37–43.
Shaughnessy, J. M. (2013). By way of introduction: Mathematics in a STEM context. *Mathematics Teaching in the Middle School, 18*(6), 324.
Smith, E. L., Parker, C. A., McKinney, D., & Grigg, J. (2018). Conditions and decisions of urban elementary teachers regarding instruction of STEM curriculum. *School Science and Mathematics, 118*, 156–168. https://doi.org/10.1111/ssm.12276
Smith, S., Roemmele, C., Miller, B. T., & Frisbee, M. D. (2018). There's something in the water. *The Science Teacher, 85*(3), 58–62.
Smith, T. M., Seshaiyer, P., Peixoto, N., Suh, J. M., Bagshaw, G., & Collins, L. K. (2013). Exploring slope with stairs and steps. *Mathematics Teaching in the Middle School, 18*(6), 371–377.
Trimble, L. (2017). Creating science websites. *The Science Teacher, 84*(9), 25–30.
U.S. Department of Education. (n.d.). *Science, technology, engineering, and math: Education for global leadership*. Retrieved from https://www.ed.gov/stem
Van den Akker, J. (2004). Curriculum perspectives: An introduction. In J. Van den Akker, W. Kuiper, & U. Hameyer (Eds.), *Curriculum landscapes and trends* (pp. 1–10). Dordrecht, The Netherlands: Kluwer Academic Publishers.
Vasquez, J. A. (2014). STEM: Beyond the acronym. *Educational Leadership, 72*(4), 10–15.
Wang, H.-H., Moore, T. J., Roehrig, G. H., & Park, M. S. (2011). STEM integration: Teacher perceptions and practice. *Journal of Pre-College Engineering Education Research, 1*(2), 1–13. https://doi.org/10.5703/1288284314636

Welling, J., & Wright, G. A. (2018). Teaching engineering design through paper rockets. *Technology and Engineering Teacher, 77*(8), 18–21.

Wenger, E. (1998). *Communities of practice: Learning, meaning, and identity*. New York: Cambridge University Press.

Wheeler, L. B., Whitworth, B. A., & Gonczi, A. L. (2014). Engineering design challenge: Building a voltaic cell in the high school chemistry classroom. *The Science Teacher, 81*(9), 30–36.

Yoshikawa, E., & Bartholomew, S. (2018). Learning technology and engineering principles through golf. *Technology and Engineering Teacher, 77*(6), 32–35.

Christopher M. Sgro is a Curriculum and Instruction Ph.D. student at the State University of New York at Albany and a chemistry teacher at Highland High School in Highland, NY. He received his Master of Arts in Teaching from the State University of New York at New Paltz (2008) and his Bachelor of Science in chemistry from the State University of New York at Binghamton (2005). His interests include integrated problem-based learning, visual representation analysis, argumentation, cognition, science classroom discourse, and eye-tracking.

Trisha Bobowski is the Principal of Kingsborough Elementary School in Gloversville, NY. She received her Master's degree in Curriculum Development and Instructional Theory from the University of Albany (2004) and her Certification of Advanced study in School District Leadership from SUNY Plattsburgh (2017). Currently, she is enrolled in the Curriculum and Instruction Ph.D. program at the University of Albany.

Alandeom W. Oliveira is an Associate Professor of science education at the State University of New York at Albany. He earned a Master's degree in science education at Southeast Missouri State University (2002) and a Ph.D. degree in science education at Indiana University Bloomington (2008). He has taught science education courses to teachers in Brazil and the US and has coordinated multiple professional development programs for school teachers, including Science Modeling for Inquiring Teachers Network, and Technology-Enhanced Multimodal Instruction in Science and Math for English Language Learners. His research interests include cooperative science learning, inquiry-based teaching, and classroom discourse.

Chapter 12
Future Elementary Teachers' Perspectives on the Importance of STEM

Lauren Madden, James E. R. Beyers, and Nicole Stanton

12.1 Introduction

The purpose of this chapter is to describe future elementary teachers' perspectives on the importance of STEM education at the elementary level. The emphasis and attention to the STEM disciplines across the K-12 spectrum over the past several decades is far reaching—from funding agencies' programs directly focused on STEM education to standards documents such as the Next Generation Science Standards (NGSS) with explicit links between science, engineering, and mathematics made clear—and is intended to engage even the youngest children in STEM learning activities. However, studies continue to show that less instructional time is devoted to science in elementary schools, in comparison to English-language arts and mathematics (e.g. Mahlzahn, 2013), and very little is known about if and how elementary teachers approach STEM in an integrated or interdisciplinary way.

This work is based on an earlier study (Madden, Beyers, & O'Brien, 2016) which used a relatively small sample (N = 73) of preservice teachers and recent graduates from the same institution to explore elementary teachers' views on the importance of STEM. The study participants responded to an online survey that included demographic information (gender, race/ethnicity, year, GPA, education major, content major) and this open-ended prompt open-ended question:

Is STEM education important at the elementary level? Why or why not?

The results of the previous study indicated that all respondents reported that STEM Education was important at the elementary level. However, a total of 10 different reasons for *why* STEM is important were listed, and the number of reasons given varied considerably across the dataset. This set of 10 responses was the basis

L. Madden (✉) · J. E. R. Beyers · N. Stanton
The College of New Jersey, Ewing, NJ, USA
e-mail: maddenl@tcnj.edu

V. L. Akerson, G. A. Buck (eds.), *Critical Questions in STEM Education*, Contemporary Trends and Issues in Science Education 51,
https://doi.org/10.1007/978-3-030-57646-2_12

for our framework for analyses, which we will refer to as *The Elementary STEM Importance Framework* (ESIF). The ESIF can be found in Table 12.1 below.

The respondents to the initial study who had content area majors that were single subject focused (e.g. mathematics, English, history) gave fewer reasons than those with more interdisciplinary majors (e.g. integrative STEM, psychology), suggesting that students with different content backgrounds had different perspectives on STEM importance.

Given the widespread adoption of the NGSS in our state and many others, and similar efforts around the globe such as the STEM Alliance initiative in the EU, and the continued disconnect between the emphasis on STEM and the way this type of instruction plays out in typical classrooms we replicated our initial study with a more robust and comprehensive approach. In this current iteration, we used a larger dataset (N = 149) in the survey, and triangulated our data by including a focus group discussion with a subset of survey respondents who were probed to clarify and refine their initial survey responses. In this current study, all participants were pre-service teachers in four or five year teacher preparation programs at our institution.

12.2 Defining STEM Education

Despite its rise in popularity, STEM education can be difficult to define or describe. This book defines STEM as a process that makes, "meaningful interdependence among all disciplines of STEM. In other words, [STEM] includes all individual disciplines of STEM (science, technology, engineering, and mathematics) in a way that is meaningful and showcases the interdependence of the fields." Even with such a comprehensive approach to STEM, it can be challenging to define who a STEM teacher is. A STEM teacher could be a person who teaches any one of the four STEM disciplines (Science, Technology, Engineering, or Mathematics), or one who teaches in a way that integrates content from two or more different content areas (Saunders, 2009). As a result the way in which teachers, current and future, discuss STEM can vary quite considerably. For example, some might emphasize the interdisciplinary nature of STEM much in the same way it is described in the *Framework for K-12 Science Education* and the NGSS (NRC, 2012, 2013). While others, such as those in Cinar, Pirasa, and Sadoglu's (2016) study of preservice science and mathematics teachers' views on STEM might see STEM as closely related to science rather than an integrated topic. Rinke, Gladstone-Brown, Kinlaw, and Cappiello (2016) found that preservice elementary teachers with methods courses explicitly tied to STEM were able to better plan integrated and interdisciplinary lessons, while those with more traditional science and mathematics methods courses were less prepared in this area. Likewise, Sumen and Calisici (2016) found that preservice elementary teachers enrolled in an environmental literacy course using a STEM perspective were better able to see environmental concepts as interdisciplinary with links between and among content areas.

Table 12.1 The Elementary STEM Importance Framework (ESIF)

Is STEM education important at the elementary level? Why or why not?	
Code	Example responses
Foundation for Later Academics	STEM education is absolutely important. An early start and understanding of STEM topics can generate greater interest. It is important to expose children to multiple fields and ideas at an early age.
Connections to Everyday Life	[STEM] helps students to understand the world around them and prepares them for real life! STEM can really help elementary school students to gain knowledge that can benefit them in the real world.
Nurturing Positive STEM Attitudes	STEM education is important that the elementary levels because it allows for students to become involved and interested in these subjects. Students can be turned off from science at a very young age and never regain that positive attitude towards it again. In elementary school, we should foster students' interests and abilities in these content areas
Integrating or Balancing Content	[STEM] not only gives students content area knowledge, but it gives them a better understanding of how everything can be related. It's nearly impossible to have one area of STEM without incorporating the other three areas as well. It's important for stem education at an elementary level to help kids begin to think strategically how science, math and technology all correlate with one another.
Preparing Students for Jobs or Replenishing the STEM Pipeline	These subjects need to be taught at [the elementary] level so the children stay in the STEM pipeline. The future is constantly changing and students need to be able to keep up with the changes.
Promotes Learning/Higher Order Thinking	STEM education also challenges children's high level thinking abilities and allows them to question everything around them (what? how? why?). Yes because it will help young students improve their problem solving skills and ability to think critically.
Promote Gender Equity in STEM	I think STEM is still seen as a strictly "boy" area of learning and we need to broaden that to all learners. I would like to see more girls in the STEM fields, breaking gender stereotypes!
Maintaining Global Competitiveness	Our [US] education [system] is falling behind and putting more emphasis on math and science is extremely important. STEM education helps students start early in learning the skills that will help our country as a whole improve in these areas and be able to compete with other countries who are currently more advanced than us.
Promote Hands-on Inquiry/Design	It is important students learn through inquiry, the design process, and exploration. [STEM] allows more "hands on" learning to be enforced rather than just memorizing.

(continued)

Table 12.1 (continued)

Is STEM education important at the elementary level? Why or why not?	
Code	Example responses
Pervasiveness of Technology	Technology is improving and advancing at a rapid rate and elementary students need to have more…experiences with STEM topics. It prepares students for a world in which everything is technologically-driven.

Note: some example responses were edited so that they addressed only the specific code identified

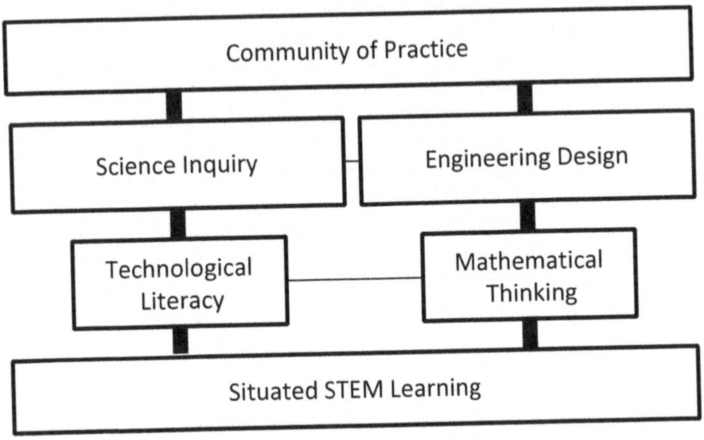

Fig. 12.1 Modified model of integrative STEM learning. (From Kelly & Knowles, 2016)

12.3 Theoretical Framework

STEM teaching and learning can take many forms, ranging from instruction in any one of the four STEM content areas to interdisciplinary approaches that use the tools and techniques from all four disciplines to design solutions. From our perspective, the integrated nature of STEM, in which science and mathematics content are used together to solve problems within a specific technological or engineering design context, is its greatest utility. Our study uses the conceptual framework developed by Kelley and Knowles (2016) which emphasizes that learning is contextual and situational and that both mathematics and technology are needed to inquire and design within a community of practice. The diagram in Fig. 12.1 below (modified from Kelley & Knowles, 2016) illustrates the interconnectedness of the STEM areas.

12.4 Methodology

Our primary objective is to better understand how prospective teachers view the *importance* of STEM education at the elementary level. A secondary objective of the current study is to evaluate the efficacy of our initial coding framework as a tool for measuring teachers' perceptions of the importance of STEM in the elementary years. Our previous study (Madden et al., 2016) suggested that preservice elementary school teachers found STEM to be important at the elementary school level for a variety of reasons, and that the number of reasons given varied considerably based on the preservice teachers' content area major.

12.4.1 Study Context

The study took place at a small, public, highly selective, primary undergraduate institution in the Northeastern United States. The population of teacher candidates in the dataset were similar to many current and future teachers at the elementary level in the US—largely female and white. The overwhelming majority of participants identified as female (n = 138 or 94%), while eight respondents (5%) identified as male, and one did not identify a gender. Most of the respondents identified as white or Caucasian (n = 123 or 84%), while the next largest groups were multiple races or ethnicities and Hispanic or Latina(o), at about 5% each (n = 8 and 7, respectively). Five participants (3%) identified as Asian, three as African American (2%), and one elected not to identify a race or ethnicity. In terms of GPA, 23 respondents (16%) elected not to respond. Of those who did respond, GPAs were almost universally greater than 3.0 with just two respondents (1%) reporting GPAs below 3.0, 15 (10%) reported GPAs between 3.0–3.24, 31 (21%) reported GPAs between 3.25–3.49, 28 (19%) reported GPAs between 3.5–3.74, and 44 (30%) reported GPAs between 3.75–4.0.

Preservice teachers across all education programs at our institution are required to enroll in a dual-major, four- or five-year program terminating in a BS or MAT. Since the focus of this study was to examine beliefs about the importance of STEM education at the elementary level, only those seeking a relevant certification were included. For example, no secondary education program students were included in this study. The relevant education major choices are listed here: Elementary Education (n = 75), Early Childhood Education (n = 20), Urban Education (at the early childhood or elementary levels; n = 1 & 8 respectively), Special Education (at the early childhood or elementary levels; n = 7 & 21, respectively), or Deaf and Hard of Hearing Education (at the early childhood and elementary levels; n = 7, 8 respectively). The dual-major content area choices include: African American Studies (n = 0), Art (n = 2), Biology (n = 3), English (n = 30), History (n = 12), integrative STEM (n = 32), Mathematics (n = 18), Music (n = 1), Psychology (n = 30), Sociology (n = 13), Spanish (n = 4), and Women, Gender, & Sexuality Studies (n = 2). Figure 12.2 below depicts the distribution of education and content area majors among respondents.

We used a mixed-methods approach in this study (Creswell, 2003) in order to reach a wide sample of future teachers and triangulate findings across multiple data sources. A single-question survey (*Is STEM education important at the elementary level? Why or why not?*) was used to determine respondents' perspectives on the importance of STEM, and all responses were coded using the ESIF. The

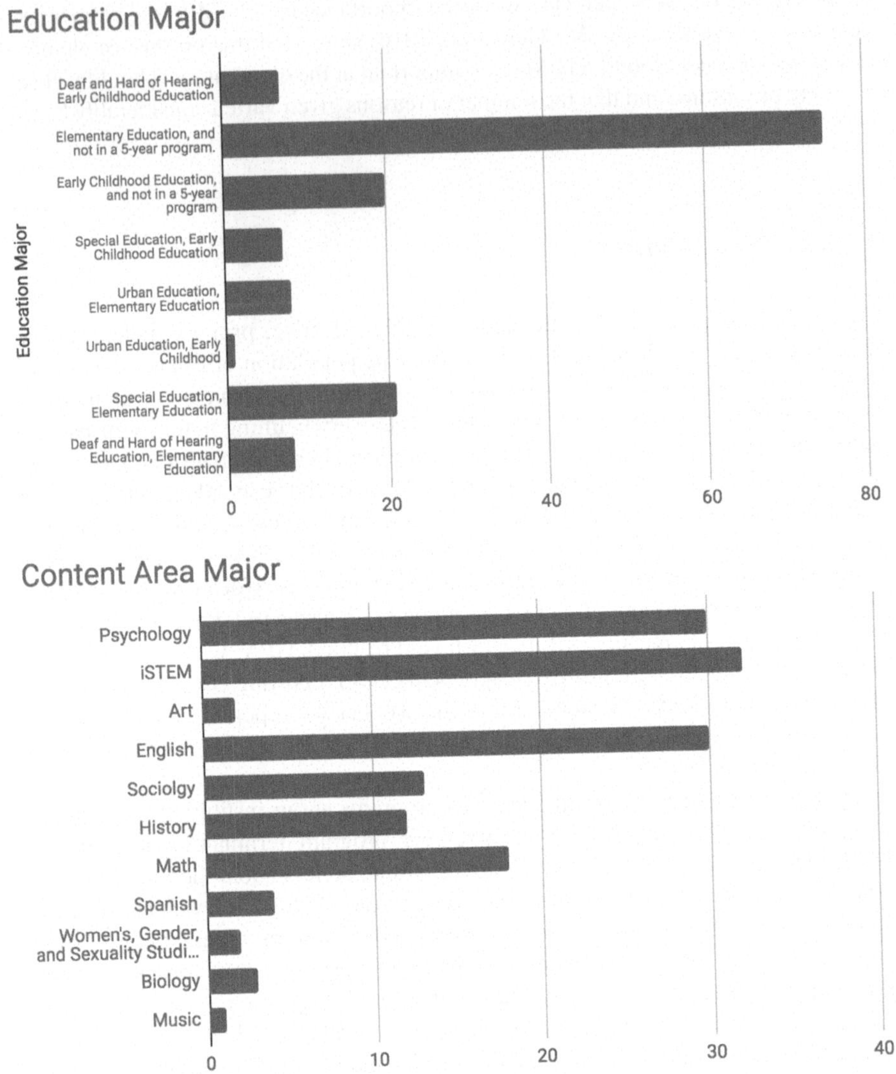

Fig. 12.2 Education and Content Area majors of participants

respondents were also asked to provide demographic information including: gender, year, GPA, education major, and content area major. The survey was distributed to all students enrolled in the four- or five-year programs described above. We distributed our initial survey to preservice teachers enrolled in teacher preparation programs at our institution and sent one follow up email to ensure that we have representation from all programs. A total of 147 responses were received.

Two coders (first and second authors) used the ESIF to analyze all reasons given for why STEM was important. To establish inter-rater reliability, the coders both coded a subset of 15 responses (just over 10%) for the presence or absence of each of the 10 codes described in the ESIF. Inter-rater agreement was 93%, and discussion took place until agreement was met on the remaining 7%. The remainder of the dataset were divided in half and coded by one coder each. Each coder spot-checked ~10% of their partner's codes and no significant discrepancies were found.

Survey respondents were also asked whether they were willing to participate in a focus group discussion to elaborate on their survey responses.[1] Focus groups were used because as Villanen (2014) noted, "Focus groups allow group interactions based on central prompts or topics to be the main data source rather than researcher-imposed questions." A total of four respondents participated in the focus group. The focus group discussion took place via Google Hangouts, and were facilitated by the third author. The facilitator asked the initial question from the survey, "Is STEM education important at the elementary level? Why or why not?" She allowed the participants to drive the interactions throughout the discussion and interjected only to ask for clarification or elaboration. The focus group discussion lasted approximately 30 min and was recorded digitally and transcribed by the facilitator. All identifying information was removed from the transcripts. The transcripts were then coded using the ESIF—instances of each of the codes that emerged in discussion were marked, and excerpts of the discussion were identified to further elaborate on participants' reasoning. We coded the transcription using the same scheme by marking examples of various codes as they emerged. It should be noted that we did not keep a count of each of the codes as it was difficult to determine which individual made which comments, but the presence or absence of each category and notable example phrases were identified. We compared responses between survey and focus group data to validate findings via data triangulation.

12.5 Findings

In our prior study (Madden et al., 2016), all respondents agreed that STEM education was important at the elementary level. During this current study, our findings in this area were similar. Nearly all respondents (143/147, 97%) reported that yes, STEM was important at the elementary level. Four (3%) reported that it was not. In the sections that follow, we will first describe the four responses which indicated that STEM was *not* important the elementary level, and then describe in detail the reasons for why STEM is important at the Elementary Level and how the frequency and distributions of those reasons given in the current study compare to those given in the previous 2016 study.

[1] For logistical reasons related to timing and scheduling during a summer session, only one focus group was held.

12.5.1 Those Who Disagree

Unlike our prior work, some responses in the current study indicate that they disagree that STEM was important at the elementary level. Each of the four respondents provided reasons in the survey for their disagreement. Two of the four commented on developmental appropriateness of STEM education; one response noted, "They're too young to understand STEM and it causes great stress on children." One respondent claimed not to know much about STEM and did not want to assert an opinion without knowing more. The fourth respondent discussed testing rather than STEM noting, "Elementary students should be taught in a more open ended way, and instead of preparing them for testing we should prepare them to be functioning adolescents." It was unclear whether this respondent understood the question based on the response given. Because the focus group participation was voluntary, we were unable to follow up with these respondents in particular.

12.5.2 Those Who Agree

In order to further explore the framework identified in previous work, we will continue with similar foci:

- Comparisons of the trends in reasons given from the current study to the previous study (Madden et al., 2016)
- Comparisons of the reasons identified rate [RIR] for reasons given by respondents in the current study compared to the previous study (Madden et al., 2016)

The percentages of overall respondents who offered a reason are shown below (see Table 12.2, below). We can see from this data that the relative frequencies for the reasons given have similar distributions insofar how the percentages are ranked. Upon further examination, though, it can be seen from the changes in the relative frequencies from study to study that there are some differences to note. The change in the percentage was calculated by dividing the two percentages and looking at the percent increase or decrease. For example, with *Foundations for Later Academics*, the percentage changed from 32 (2016) to 38.5 (2018). Dividing these two values gives the approximate result of 1.20. This means that there has been a 20% increase from 2016 to 2018 in the relative frequency for this code. Across the board, there were increases in six of the ten percentages representing the relative frequency of the coded reasons. Interestingly, five of the six increases were 19% or greater. The decreases in the relative frequency were, by comparison, modest, with the exception of *Nurturing a Positive STEM Attitude*. This category decreased by 15.4%.

Additionally, it can be seen in Table 12.3 (see Table 12.3, below) that there were also some substantial changes in the Reasons Identified Rate [RIR]. This rate is calculated as a ratio of the number of reasons identified per respondent in each of the respective categories below, e.g., content major. Higher RIRs indicate that the

Table 12.2 Reasons given by participants for why STEM is important

Reasons (codes)	Percentage including this Reason (2018)	Percentage including this Reason (2016)	Change in Relative Frequency from 2016 to 2018
Foundation for Later Academics	38.5%	32 %	Up 20%
Connections to Everyday Life	27.27%	26 %	Up 4.8%
Nurturing Positive STEM Attitudes	21.67%	25 %	Down 15.4%
Integrating or Balancing Content	21.67%	22 %	Down 1%
Preparing Students for STEM Jobs	20.97%	22%	Down 5%
Promotes Learning/ Higher Order Thinking	25.17%	21%	Up 19%
Pervasiveness of Technology	20.97%	14%	Up 49.7%
Promote Hands-on Inquiry/Design	13.28%	11%	Up 20.7%
Promote Gender Equity in STEM	5.59%	4%	Up 39.75%
Maintaining Global Competitiveness	2.79%	3%	Down 7.5%

response included multiple reasons, thus providing a more nuanced explanation for why STEM is important. The RIR for the 2016 study is given next to the RIR for the current study.

The largest increases in RIR were among Mathematics, English, History, Sociology, and Women, Gender, & Sexuality Studies majors, at up 27%, 22.7%, 28%, 9.2% and 757% respectively. There was little or no change among the remaining content majors for whom a change could be calculated. Also, the overall RIR increased by 12.9% from 1.78 in 2016 to 2.01 in the current study.

In order to develop a better understanding about the reported beliefs about STEM education at the elementary level, we sought additional data from respondents in a follow-up focus group discussion.

12.5.3 Focus Group Discussion Trends

The individuals who agreed to participate in the focus group unanimously agreed that STEM was important at the elementary level. During the discussion, seven of the 10 coding categories from the ESIF were addressed. These were: *Foundation for Later Academics, Connections to Everyday Life, Nurturing Positive STEM Attitudes, Integrating or Balancing Content, Preparing Students for Jobs or Replenishing the*

Table 12.3 Reasons identified rate [RIR][a]

Second Major	RIR by content major (2018)	RIR by content major (2016)	Percent Change in RIR
Integrative-STEM (n = 31)	2.03	2.08	Up 2%
Psychology (n = 28)	2.11	2.07	Up 2%
Spanish (n = 4)	2	2	No change
Mathematics (n = 17)	2.18	1.71	Up 27%
English (n = 29)	2	1.63	Up 22.7%
History (n = 12)	1.83	1.43	Up 28%
Sociology (n = 12)	1.53	1.40	Up 9.2%
Biology (n = 3)	2	0[b]	N/A[b]
Art (n = 2)	1.5	0[b]	N/A[b]
Music (n = 1)	2	0[b]	N/A[b]
Women, Gender, & Sexuality Studies (n = 2)	2.5	.33	Up 757%
Overall RIR	**2.01**	**1.78**	**Up 12.9%**

[a]For the purposes of these analyses, the reasons identified rate (RIR) is operationalized as the rate of reasons given for why STEM education is important per respondent overall (overall RIR), and per respondent within a particular second major designation (e.g., the number of reasons given by history majors divided by the total number history majors). Examples of the calculations are given in the text that follows the initial description

[b]No percent change in RIR was calculated here because the RIR in 2016 was 0. For these content majors, it was because there was no representation from that content area

STEM Pipeline, *Promotes Learning or Higher-Order Thinking*, and *Promotes Gender Equality in STEM*. Three codes did not emerge as part of the conversation during the focus group. These were: *Maintaining Global Competitiveness, Promotes Hands-on Inquiry or Design, and the Pervasiveness of Technology*. Table 12.4 lists examples of codes that emerged during the focus group discussion.

12.6 Discussion

When we look at the data collected during the present study, we found many similarities to our prior work that go beyond simply agreeing that STEM is important at the elementary level. For example, the category *Foundation for Later Academics* continues to be the most frequently cited reason, even more so during this iteration, with a 20% increase in response rate. This code came up quite frequently in the focus group discussion as well, as one participant noted, “In elementary school you have the building blocks of all the education you’re going to do in the future, your like high school and middle school. So, introducing iSTEM [*sic*] Education as early as possible, I just think, is super important.” Interestingly, this focus on the early years contrasts the two responses who disagreed with the idea that STEM was important at the elementary level. The *Connection to Everyday Life* also remained a popular reason given by respondents, though the relative frequency percentage decreased slightly from our earlier study, down about 5%. Again, this response

Table 12.4 Example ESIF categories cited during the focus group discussion

Code	Example responses
Foundation for Later Academics	In elementary school you have the building blocks of all the education you're going to do in the future, like in high school and middle school. So, introducing iSTEM Education as early as possible, I just think, is super important
Connections to Everyday Life	It's kind of teaching people that these things are fundamental in your use and in your everyday life. It's not just fun for kids.
Nurturing Positive STEM Attitudes	When you're young you kind of get the association that math is really bad, or that it's really hard, it's impossible, I'm never going to use this, and that's really just not true. So, if we build the foundation at a younger age, maybe that stigma will sort of go away.
Integrating or Balancing Content	Doing science experiments combines math and science and then connecting math and science and not just making them isolated subjects. This [content] all comes together and has use for us.
Preparing Students for Jobs or Replenishing the STEM Pipeline	STEM is used in all areas you know, it's not just for the future, like all jobs involve STEM in some way, shape, or form. It's important for the young students to build a really good foundation for so then as they go through school, get better and better at it, and then use it, like actually use it in the future.
Promotes Learning/ Higher Order Thinking	I think STEM is basically that you gain problem solving skills and no matter what you do in real life, you are going to have to solve problems"
Promote Gender Equity in STEM	You see the gender gap between people in the field of STEM. So hitting students early on saying you can do science, you can do math, you can be a woman in engineering, the sooner you start making sure that kids understand gender bias, you will have more women in STEM.

Note: some example responses were edited for clarity

came through in the focus group discussion also, as one participant noted, "[Elementary STEM is] kind of teaching people that these things are fundamental in your use and in your everyday life. It's not just fun for kids."

The next most frequently given reason on survey responses was *Promotes Learning or Higher Order Thinking*. Though this code was prevalent during our earlier study, it was cited by participants nearly 20% more during the current study, and focus group discussions echoed they found this important. In the words of one participant: "I think STEM is basically that you gain problem solving skills and no matter what you do in real life, you are going to have to solve problems." Though it remained frequently-given reason for STEM importance, *Nurturing Positive STEM Attitudes* decreased in the current survey responses by about 15%, and it was mentioned by focus group participants as well, in some cases with specific references to their own learning experiences, "And I think that even if it's not confidence, if it's enthusiasm about the subject because I've had math teachers tell me that it's ok that I'm not good at math, you don't have to be good at everything. But I was like...there has to be some level of that's not ok for a teacher to say to a student who is struggling. Instead, there should be some sort of encouragement and things there so even if a teacher isn't confident (...) they can kind of encourage and say I'm also working on it."

Though the overall response rate was low for *Pervasiveness of Technology*, it did increase by nearly 50% across the two administrations, but did not come up during the focus group discussion. The percentage of survey responses that discussed *Preparing Students for STEM Jobs* and *Integrating or Balancing Content* remained about the same between the two administrations of the survey, and these categories were also addressed during the focus group discussion. The two lowest categories of responses *Promotes Gender Equity* and the *Maintaining Global Competitiveness* remained the lowest during both survey administrations. It should be noted that the attention to gender equity increased between administrations and was a key component of the focus group discussions. This attention yielded several responses from students that connected to their personal experiences in higher education. As one said, "I can't tell you how many times my friends have come up to me and they're like, 'My own professor forgot I was a student in their class because I was a woman, sitting, waiting for an engineering lecture to start that was like extra on a Wednesday afternoon and [the professors] were like 'Actually, we need this room for engineering whatever' and [we]'re like, 'we're all engineers.' Like that kind of 'who can be involved in STEM' is just as important as the practice of STEM itself."

Another interesting comparison between the two administrations of the ESIF study was the increase in the reasons identified rate (RIR) from 1.79 in the 2016 study to 2.01 currently. This suggests that perhaps that there was some shift in the complexity in the ways individuals in our program thought about STEM education in the elementary years. Interestingly, there were also some shifts in RIR when we consider content area majors. In our 2016 study, the participants majoring in subjects that considered multiple perspectives, namely integrative STEM and psychology tended to identify more reasons in their explanations for why STEM is important, while single-subject majors such as English and mathematics mentioned fewer. In the current study, however, integrative STEM and Psychology majors continued to offer among the higher high RIRs, but English, History, Mathematics, and Sociology, all showed strong increases in RIR, ranging from 9–28%, with Mathematics majors taking the lead this time, with an RIR of 2.18. The 2018 administration had respondents from a greater number of majors, adding Art, Biology, and Music to the list, so comparisons cannot be made in these areas. It should be noted that the Women, Gender, and Sexualities content major increased by 737%, but this was due largely in part to a very low initial RIR (.33) from in the 2016 study. For this reason, it is being considered an outlier.

These shifts in RIR overall and within individual majors suggests that perhaps there are some shifts in the types of experiences individuals encounter in our teacher preparation program, or in society overall. Other studies (e.g. Cinar et al., 2016; Rinke et al., 2016; Sümen & Çalisici, 2016) support the idea that preservice elementary teachers' views on STEM can be shaped by coursework. In mathematics education literature, it is similarly suggested that dispositional beliefs can vary according to content topics (Beyers, 2005), which in turn can impact the ways in which students engage in learning opportunities in that content (Beyers, 2011, 2012). The implication here being that it is conceivable to suggest that the changes in RIR may be a reflection of changes in content learning experiences of students in their teacher preparation program. We are aware that there is a more intense focus on the structure and use of the NGSS in the science methods courses offered at our institution, and this could also contribute to the overall increases in RIR. Additionally, one focus group participant referenced female faculty members by name when discussing gender equity. These comments suggest that perhaps course content has changed to include discussion of these issues more broadly at our institution. The RIR increases may also be due, in part, to reaching a larger and more representative sample of future elementary education majors at our institution. Regardless of reason, when we consider our adapted version of Kelly and Knowles' (2016) model for integrated STEM education focused on problem solving within the context of a greater community of practice, an increase in the number of reasons given for STEM importance supports this model.

12.7 Conclusions

In sum, we found that there continues to be large scale agreement in the value and importance of Elementary STEM among preservice teachers at our institution, and a broad variety of reasons given for this importance ranging from considering foundational knowledge of students to bigger scale issues such as equity and career readiness of future students. This range of reasons, and increase in complexity in the number of reasons reported by students suggests that preservice teachers at our institution think of STEM as a multi-faceted discipline.

We can also say with confidence that the ESIF served as a valuable and reliable tool for describing reasons why STEM is important among preservice elementary teachers. The tool was useful for analyzing responses to survey data as well as transcribed focus group interviews, and provides researchers with a model for framing discussions of STEM importance. It also allowed us to quantify some of the ways in which the views of preservice teachers at our institution have shifted with regard to the importance of STEM.

12.8 Implications and Future Work

Our findings, coupled with the literature (e.g Cinar et al., 2016; Rinke et al., 2016; Sümen & Çalisici, 2016) suggest that elementary teachers' views of the importance of STEM education at the elementary level can be complex, though there are clearly some students who are not convinced, as evidenced by the four responses to this current survey. This implies that shifts in coursework for preservice teachers to directly address the reasons for STEM importance could be fruitful. Though the respondents were not asked to identify reasons for their beliefs (i.e. what experiences or knowledge did they draw on when responding to the survey) future administrations of the ESIF could include probing questions to help parse out ways in which courses might already be addressing this area or could do so more explicitly.

The study demonstrated the utility of the ESIF as an analytical framework. Future analyses of this dataset could be used to identify differences among other factors, such as gender or education major, to tell us a bit more about differences in the reasons given for STEM importance. The survey could also be administered to practicing teachers, college faculty, or elementary aged children to inform us on how these audiences differ in their views of STEM and inform future professional development efforts.

References

Beyers, J. E. R. (2005). What counts as "productive" dispositions among pre-service teachers? In G. M. Lloyd, M. R. Wilson, J. L. M. Wilkins, &; S. L. Behm (Eds.), *Proceedings of the 27th annual meeting of the North American Chapter of the International Group for the Psychology of Mathematics Education*. Retrieved from http://convention2.allacademic.com/index.php?cmd=pmena_guest [Publisher: PME-NA]

Beyers, J. E. R. (2011). Student dispositions with respect to mathematics: What the current literature says. Chapter 5 in the 73rd yearbook of NCTM (2011). In *Motivation and disposition: Pathways to learning mathematics*. Reston, VA: National Council of Teachers of Mathematics.

Beyers, J. E. R. (2012). *An examination of the relationship between prospective teachers' dispositions and achievement in a mathematics content course for elementary education majors*. SAGE Open, October–December 2012; vol. 2, 4: 2158244012462589.

Cinar, S., Pirasa, N., & Sadoglu, G. P. (2016). Views of science and mathematics pre-service teachers regarding STEM. *Universal Journal of Educational Research, 4*(6), 1479–1487.

Creswell, J. W. (2003). *Qualitative, quantitative, and mixed methods approaches*. Sage.

Kelley, T. R., & Knowles, J. G. (2016). A conceptual framework for integrated STEM education. *International Journal of STEM Education, 3*(11). https://doi.org/10.1186/s40594-016-0046-z

Madden, L., Beyers, J., & O'Brien, S. (2016). The importance of STEM education in the elementary grades: Learning from pre-service and novice teachers' perspectives. *Electronic Journal of Science Education, 20*(5), 1–18.

Malzahn, K. A. (2013, September). *2012 National survey of science and mathematics education-status of elementary school mathematics teaching*. Retrieved on October 9, 2013 from: http://www.horizon-research.com/reports/?sort=report_category

National Research Council. (2012). *A framework for K-12 science education: Practices, crosscutting concepts, and core ideas*. Washington, DC: National Academies Press.

National Research Council. (2013). *Next generation science standards: For states, by states*. Washington, DC: National Academies Press. Retrieved from: nextgenscience.org

Rinke, C. R., Gladstone-Brown, W., Kinlaw, C. R., & Cappiello, J. (2016). Characterizing STEM teacher education: Affordances and constraints of explicit STEM preparation for elementary teachers. *School Science and Mathematics, 116*(6), 300–309.

Sanders, M. (2009). STEM, STEM education, STEMmania, *The Technology Teacher,* Dec./Jan.-2009, 20–26.

Sümen, Ö. Ö., & Çalisici, H. (2016). Pre-service teachers' mind maps and opinions on STEM education implemented in an environmental literacy course. *Educational sciences: Theory and practice, 16*(2), 459–476.

Villanen, H. (2014). Teachers' reflections on an education for sustainable development project. *International Research in Geographical and Environmental Education, 23*(2), 179–191.

Lauren Madden is an Associate Professor of Elementary Science Education and Coordinator of the Environmental Sustainability Education Minor in the Department of Elementary and Early Childhood Education at The College of New Jersey. Her teaching and research focus on advocating for scientific literacy and the health of our planet through teaching and learning.

James E. R. Beyers is an Associate Professor of Mathematics Education and Department Chair for Elementary and Early Childhood Education at The College of New Jersey. His research focuses on prospective teachers' mathematical dispositions and their beliefs about the importance of STEM Education at the elementary school level. He is the current President of the New Jersey Association of Mathematics Teacher Educators.

Nicole Stanton is a current student at The College of New Jersey studying Special Education and iSTEM education. She will graduate in May 2020 with a Bachelor's in iSTEM education and in May 2021 with a Master's in Special Education. She hopes to be able to instill a love of STEM in her future students.

Chapter 13
Switching Lanes or Exiting? STEM Experiences, Perceptions, and Identity Construction Among College STEM Switchers

Youngjin Song, Ann Y. Kim, Lisa M. Martin-Hansen, and Elaine Villanueva Bernal

13.1 Introduction

A larger proportion of students leave their initial STEM majors than that of other disciplines, although nearly 40% freshmen plan to major in STEM and the number of undergraduate students enrolled in STEM disciplines has increased in the last 10 years in the United States (National Academies of Sciences, Engineering, and Medicine, 2016). This is concerning for many reasons, including the facts that the Bureau of Labor Statistics reports that there is above-average growth of STEM occupations compared to other sectors (Fayer, Lacey, & Watson, 2017), and there are labor shortages in some STEM industries (Xue & Larson, 2015). Scholars have investigated various factors to explain why students switch out of their initial STEM majors, often called *STEM switchers* (e.g., Chen, 2013; Seymour & Hewitt, 1997). However, a question unanswered in many of these STEM switching studies is where STEM switchers go and why or how they choose their new majors.

Understanding why STEM students switch as well as an examination of where students go—and for what reasons—can shed light on the phenomenon of STEM switching. For example, one student may switch from a STEM field to a non-STEM field because of issues with instructional delivery. In contrast, another student may switch from a STEM field to another STEM field because s/he develops a specific interest while taking elective courses. These students have different needs. It is

Y. Song (✉) · L. M. Martin-Hansen
Department of Science Education, California State University, Long Beach, CA, USA
e-mail: youngjin.song@csulb.edu

A. Y. Kim
Department of Human Development, California State University, Long Beach, CA, USA

E. V. Bernal
Department of Chemistry and Biochemistry, California State University, Long Beach, CA, USA

V. L. Akerson, G. A. Buck (eds.), *Critical Questions in STEM Education*, Contemporary Trends and Issues in Science Education 51,
https://doi.org/10.1007/978-3-030-57646-2_13

necessary to comprehend the whole process of students' STEM switching so higher education institutions can learn how to improve student learning experiences or to effectively advise on the choice of a career.

To gain a better understanding of the process of STEM switchers' final destination choice—whether it is STEM or non-STEM—out of their initial STEM majors, our work is grounded in the theoretical framework of students' identity. Several researchers have studied how students' identity and their entry choice of studying STEM in higher education are interrelated (e.g., Aschbacher, Li, & Roth, 2010; Holmegaard, Madsen, & Ulriksen, 2014). Research has yet to capture the understanding of students' second choice of study after they switch out of STEM majors with the lens of identity at the college level. Thus, this chapter contributes to the study of why STEM switchers choose other majors by analyzing their experiences in and out of STEM fields and their choice as a process of identity construction. Moreover, previous STEM switching research has not been able to capture how STEM switchers characterize their understanding of STEM in terms of how the disciplines (S/T/E/M) are interrelated within a major or career. This is important as students switching from their initial STEM majors are referred to as 'STEM switchers' as opposed to 'chemistry switchers' or 'engineering switchers.' Holmegaard et al. (2014) pointed out high school students' perceptions of STEM together with their identity affected their inclination to study STEM fields in higher education institutions. Thus, this chapter seeks out plausible explanations of how STEM switchers' perceptions of STEM related to their choices of switching.

The following research questions guided our work:

With the college students who switched out of their initial STEM majors,

1. How do students' STEM learning experiences relate to their STEM identity narrative and choices of major switching?
2. How do students' perceptions of STEM disciplines relate to their choices of major switching?

13.2 Literature Review

13.2.1 What Is STEM? The Interdependent Nature of STEM

Currently, science, technology, engineering, and mathematics are referred to together as STEM but there is little agreement among groups and scholars about the definition of STEM and little acknowledgement of why those disciplines should be grouped together (Breiner, Johnson, Harkness, & Koehler, 2012; Chesky & Wolfmeyer, 2015; Williams, 2011). On one hand, there is a common idea of STEM that science, technology, engineering, and mathematics are distinct fields that have yet to be integrated. This is particularly apparent when examining documents from different government entities in the United States. The National Science Foundation has a four-page list of STEM fields. The National Academies simply provide a bulleted-description for each of the letters S/T/E/M (National Academies of Sciences, Engineering, and Medicine, 2016). On the other hand, STEM is treated as one discipline in and of itself so that there is a nature of STEM (e.g., "NOSTEM"; Peter-Burton, 2014), especially in the discourse of research regarding STEM switching.

Our team posits that fields of basic sciences (e.g., biology, chemistry, physics, and earth sciences), applied sciences (e.g., agriculture, environmental sciences, and forensic sciences), technology (e.g., computer, internet, and satellites), and engineering (e.g., chemical, civil, and mechanical) are distinct but meaningfully interdependent as they build on one another. For instance, basic sciences borrow from each other and the disciplines do not function independently. pH is used to describe biological systems and logarithmic scales are used to describe pH (Tro, 2011). Molecular dynamics goes hand in hand with quantum physics and mathematics (Atkins, 2018). Materials science and engineering, which involves the discovery, design, and development of new or existing materials, uses principles of biology, chemistry, and physics to study the nature of materials (Callister & Rethwisch, 2014), and all of these developments are with the help of technology (Dogan & Robin, 2015). Computational biology and chemistry use computers and mathematical modeling to characterize ecological systems and molecular structures, and quantum computing has enabled rapid design and development of solid-state batteries. Computer Aided Design (CAD) is ubiquitous across engineering disciplines. In these ways we recognize that STEM disciplines inform and rely upon one another (Akerson et al., 2018).

In college, students are more likely to experience STEM as compartmentalized disciplines. Higher education institutions typically make distinctions regarding what is considered STEM and non-STEM; there are STEM and non-STEM colleges and majors. As a starting point of our investigation on STEM switching patterns, our research team stayed close to the way that our institution considered STEM vs. non-STEM. We categorized STEM majors as those housed in two colleges—the College of Engineering (COE) and the College of Natural Sciences and Mathematics (CNSM). This made it possible to track within the university system where students switched to. In the process of our investigation we saw other fields of study that may be considered STEM at other institutions, such as health science and industrial design, which were termed by our research team as *STEM-adjacent*. Students in *STEM-adjacent* majors such as kinesiology, industrial design, or economics often learn STEM content to apply it to their majors (all STEM-adjacent majors had three or more required STEM courses). The category *non-STEM* designates majors requiring less than three science, technology, engineering or mathematics courses and outside of the COE and CNSM.

13.2.2 Students' STEM Experiences

Concerns about college students switching out of STEM fields is not new. Seymour and Hewitt (1997) comprehensively explored the wide range of reasons for students' decision of switching out of the Science, Mathematics, and Engineering (SME; the acronym used in the book) majors. Based on a three-year ethnographic study, the authors categorized and ranked 23 factors contributing to SME switching decisions. Among those, the top four factors were: "lack or loss of interest in science; belief that a non-SME major holds more interest, or offers a better education; poor teaching by SME faculty; and feeling overwhelmed by the pace and load of curriculum demands" (p. 32). Seymour and Hewitt found institutional structure and culture played a key role in SME switching, which challenged the common belief that most switching was caused only by individual student deficiencies. They also presented a set of concerns all SME students commonly experienced, indicating there were more similarities between switchers and non-switchers with regards to "abilities, motivations, and study-related behaviors" (p. 30).

Researchers have continued to investigate the various factors that cause students to switch out of their initial STEM majors (e.g., Chen, 2013; National Academies of Sciences, Engineering, and Medicine, 2016). Some important factors are directly related to college STEM courses including faculty pedagogy (e.g., Xu, 2018), negative experiences in introductory classes that functioned as a gatekeeper or gateway (e.g., Crisp, Nora, & Taggart, 2009; Gasiewski, Eagan, Garcia, Hurtado, & Chang, 2012; Mervis, 2010), and academic performance. In addition, researchers have paid attention to contextual factors such as academic support programs (e.g., advising/counseling), interaction with (including support from) faculty and peers, and university structural features (e.g., class size). Several attempts have been made to improve academic support programs that resulted in greater retention in STEM majors (e.g., Polnariev et al., 2017; Wilson et al., 2012). Commonly these programs built students' connections to communities in their early STEM careers so they would receive needed support from faculty and peers.

Other explanations related to students' individual attributes such as gender, ethnicity, and family backgrounds have been examined, with particular concerns about the persistent underrepresentation of women and racial/ethnic minority students (National Academies of Sciences, Engineering, and Medicine, 2016). Some find they leave STEM fields at higher rates than their male counterparts (Astorne-Figari & Speer, 2018; Green & Sanderson, 2018; Maltese & Cooper, 2017). However, recent research studies on racial/ethnic minority students' major switching found that students' ethnicity was not a significant predictor of their intention to drop out of or change STEM majors (Shedlosky-Shoemaker & Fautch, 2015; Thompson & Bolin, 2011; Xu, 2018). Additionally, STEM entrants' parental education, income levels, and demographic characteristics were not significantly associated with the outcome of leaving STEM fields by switching majors (Chen, 2013).

Despite identifying factors that contribute to student switching, the phenomenon still exists. Therefore, in the current study, we propose investigating the experiences of STEM entrants, the meaning put on those experiences, and other college experiences informing where they go as important sources of information to gain insight into the phenomenon. Particularly, we apply an identity lens to understand STEM students' decision of switching out of initial STEM majors.

13.2.3 STEM Identity

Identity has been identified as a valuable way to examine student retention, persistence, achievement, graduation from STEM fields (e.g., Aschbacher et al., 2010; Carlone & Johnson, 2007; Nadelson et al., 2017; Shanahan, 2009). Scholars have been interested in which identity facets keep students in STEM as well as which push them out. We expand and build on the science identity theoretical frameworks by Carlone and Johnson (2007) and by Kim and Sinatra (2018; interactionist framework) and apply them to student experiences in various STEM fields. Carlone and Johnson (2007) identified competence, performance, and recognition as the three key areas making up one's science identity. In other words, a student's science identity is informed by her knowledge of science and competence in science spaces, how she demonstrates or performs this competence, and herself recognition and recognition of others as a "science person." In their interactionist framework, Kim and Sinatra (2018) argued that science identity development is informed by experiences that either encourage individuals to approach or avoid science. Both frameworks focus on general, universal experiences, feelings, and thoughts. Therefore, the frameworks were considered appropriate for understanding student experiences in STEM that included specific academic disciplines and beyond.

Since the initial publication of Carlone and Johnson's (2007) framework, other scholars have contributed to its development, modification, and expansion. Hazari et al. (2010) limited the focus of recognition to that from others and emphasized the importance of a student's interest in physics when understanding a student's physics identity. In a case study, Mark (2017) presented an African American, male high school student Randy who wanted to pursue a degree in engineering at MIT. Randy presented a STEM identity that included his interest in STEM-related topics (e.g., problem-solving, etc.), his newly acquired STEM content knowledge, and also his interest in making a lot of money (Randy refers to himself as "the investor" p. 995). Mark referred to this as the student's "economic lens."

Additional research supports broadening what STEM identity encompasses beyond the domains of competence, performance, recognition, and interest to consider how it is informed. In a recent literature review of middle and high school female students' STEM experiences, Kim, Sinatra, and Seyranian (2018) emphasized the role of the environment. In their review, the environment included friends/peers, family, teachers, as well as curriculum and classroom environment. Students experienced messages from their environment whether or not they should be pursuing STEM majors and careers. Attention to the environment and background of students at the postsecondary level is emphasized by work from McGee (2016). In her interview study of Black and Latinx STEM students she identified how challenging these racialized STEM spaces can be. Students were putting in additional psychological work to combat race-ethnicity stereotypes and deal with psychological exhaustion.

In addition to such focused STEM identities scholars have investigated the role of career-related motivations to be persistent in STEM or using STEM interest to inform career decisions. Some scholars have focused on the exploration of future careers and its relation to STEM persistence (Perez, Cromley, & Kaplan, 2014) and applied social cognitive career theory in understanding STEM learning (Nugent et al., 2015) and career pursuits (Byars-Winston, 2014). Others have focused on recruiting STEM-talented students (Lee & Nason, 2013) or STEM professionals to become teachers (Grier & Johnston, 2009; Snyder, Oliveira, & Paska, 2013) and developing a teacher identity. Given this research is on STEM students' and professionals' career choices and career identities, we consider STEM identity to include not only competence, performance, recognition, and interest in one STEM domain, but also one's experiences in other STEM domains as well as future career goals. Considering the interdependent nature of STEM fields, we question whether students leaving one STEM field consider other STEM fields of study as it keeps their STEM identity intact.

13.3 Method

13.3.1 Research Setting

The study took place at the 4-year public university located in the west coast region of the US during the 2018–2019 academic year. Thus, we acknowledge that our study represented only a U.S. perspective in a particular context. The university has eight colleges with a student body of about 37,000. As a Hispanic-Serving Institution (HSI) and Asian American, Native American, and Pacific Islander-Serving Institution (AANAPISI), the university is one of the west coast's top universities in terms of students' ethnic diversity. The demographics of the campus are approximately 40% Hispanic/Latinx, 20% Asian, 20% White, <10% non-resident alien, <5% Black of African American, and <1% Native Hawaiian or Other Pacific Islander, American Indian or Alaska Native. To investigate students' STEM switching, 18 majors were considered as STEM majors in the two colleges: CNSM[1] and COE.[2]

[1] Seven majors exist in the CNSM: Biological Sciences, Chemistry and Biochemistry, Environmental Science and Policy, Geological Sciences, Mathematics (math education is included here) and Statistics, Physics and Astronomy, and Science Education.

[2] The COE offers undergraduate degrees in 11 different disciplines: Aerospace Engineering, Biomedical Engineering, Chemical Engineering, Civil Engineering, Computer Engineering,

13.3.2 Participants and Data Collection

Participants came from diverse ethnic backgrounds and all had some or all science learning experiences in U.S. middle and high schools. Table 13.1 summarizes participants' information including their initial major, semester at which they switched, final major, gender, and ethnicity. The participants were described with pseudonyms to protect their confidentiality. In our findings, we briefly presented participants' information in parentheses, using the ethnic-racial labels and gender pronouns given to us by the participants. Eight students switched from STEM to another STEM majors (e.g., Biomedical Engineering to Biochemistry), seven students switched to STEM-adjacent majors (e.g., Chemistry to Health Science), and eight students switched to non-STEM majors (e.g., Biology to Psychology). In our findings, we also labeled the switching patterns with abbreviations in parentheses with SS referring to a STEM to STEM switch, SS-adj referring to a STEM to STEM-adjacent switch, and SN referring to a STEM to non-STEM switch.

Table 13.1 Participant information

Pseudonym	Gender	Ethnicity	Starting year	Initial major	Switching year	Final (or current) major
STEM to STEM switchers						
Kelsey	Female	White	2015 Fall	Physics	2017 Fall	Math/Physics
Kirsty	Female	White	2017 Fall	Marine Biology	2018 Spring	Math Education
Danae	Female	Asian	2015 Fall	Marine Biology	2018 Spring	Environmental Science
Lily	Female	Asian	2017 Fall	Biology	2018 Fall	Biochemistry
Chloe	Female	Latinx	2016 Fall	Biochemistry	2017 Spring	Molecular cell bio
Gaby	Female	Latinx/ Indian	2015 Fall	Mechanical Engineering	2018 Summer	Physics
Daniel	Male	White	2013 Fall	Physics	2015 Fall 2017 Fall	Chemical Engineering Biochemistry
Joshua	Male	Asian/ Pacific Islander	2013 Fall	Biomedical Engineering	2015 Spring	Biochemistry
STEM to STEM-adjacent switchers						
Addie	Female	White	2016 Fall	Biology	2017 Spring	Industrial Design

(continued)

Computer Science, Construction Engineering Management, Electrical Engineering, Electronics Engineering Technology, Computer Engineering Technology, and Mechanical Engineering.

Table 13.1 (continued)

Pseudonym	Gender	Ethnicity	Starting year	Initial major	Switching year	Final (or current) major
Jacee	Female	Asian	2015 Fall	Pre-Chemistry	2017 Fall	Health Science
Alicia	Female	Asian	2015 Fall	Computer Science	2017 Spring	Information Systems
Sarah	Female	Latinx	2013 Fall	Biology	2014 Spring	Health Science
Jared	Male	White	2014 Fall	Mech Engineering	2015 Fall	Economics
Jeremy	Male	Asian	2015 Fall	Computer Engineering	2017 Spring	Information Systems
Calvin	Male	Half Asian; Half White	2013 Fall	Mechanical Engineering	2014 Spring	Industrial Design
STEM to non-STEM switchers						
Mary	Female	White	2014 Fall	Biology	2015 Spring	Psychology
Hannah	Female	Asian	2016 Fall	Chemistry	2018 Spring	Recreation
Sofia	Female	Asian	2015 Fall	Biology	2017 Fall	Psychology
Angela	Female	Pacific Islander	2015 Fall	Biology	2017 Spring	Psychology
Tonya	Female	Latinx	2017 Fall	Aerospace Engineering	2018 Fall	Film
Chris	Male	Asian/ European	2013 Fall	Organismal Biology	2015 Spring	Psychology
Carl	Male	Asian/ Pacific Islander	2013 Fall	Biology	2014 Fall	Communication
Marco	Male	Latinx	2016 Fall	Pre-Mech Engineering	2018 Spring	Consumer Affairs

After IRB approval, the university provided our research team with a list of individual students who initially enrolled in a STEM major (from 2014 to 2017) but had switched out of these majors. This list consisted of 1634 students who started with a STEM discipline. The Advising Office of CNSM sent out interview invitation emails to this list of students several times. Those who were interested in participating voluntarily replied back to one of the authors. Participants were asked to choose between in-person or via video call (e.g., Skype, Zoom, etc.) interview. Initial response was slow and so we extended the time window for participant recruitment as long as possible. In the end we were able to interview a total of 23 participants (15 females). Each received a $25 gift card for participation.

Interviews were conducted in the manner of semi-structured conversations. Using Seymour and Hewett's (1997) interview protocol as a foundation we added questions to reflect the elements of STEM identity and to probe participants' perceptions of STEM related to their major switching. We asked participants: their interest and decision-making process for choosing the initial STEM majors, for switching majors, and the environment around them (e.g., Who and what influenced your decision? How did you pick the major to switch to? What did you consider?); perceptions of their performance and competencies as STEM and non-STEM majors (e.g., Could you identify successes/challenges that you faced in taking STEM courses?); and their images about STEM fields/disciplines and STEM careers and how they fit in with those images (e.g., What was the image you had of STEM majors and STEM careers initially and now?). The full interview protocol is included in Appendix. All interviews lasted 20–45 min, were audio recorded, and transcribed verbatim by a separate transcriber.

13.3.3 Data Analysis

Interview data were analyzed using thematic analysis (Braun & Clarke, 2006) after the interviews were transcribed. We first open-coded using line-by-line analysis (Strauss & Corbin, 1990) without preconceived codes. We used both *in vivo* codes and researcher codes in generating many possible open codes in the transcripts. Each author individually coded two transcripts and when needed, listened to the interviews to catch a subtle nuance or to correct any errors in the transcripts. Next, two authors met with their open codes and generated a list of initial codes. The remaining two authors also read the same transcripts and added, modified, and expanded on those initial codes till consensus was reached. Then, we coded the remaining data using the agreed codes and organized them into categories and themes. All coding categories, patterns, and themes from data were constantly compared to each other (Glaser & Strauss, 1967). During this process, we regularly discussed differences in interpretation as a group until a consensus was reached (no exact correlation figure is stated as did not use a software package for qualitative analysis) for investigator triangulation, which built the credibility of our qualitative research (Denzin, 2017).

13.4 Results

13.4.1 STEM Experiences, STEM Identity, and Switching Majors

Strong, Positive STEM Identity at Start Twenty of the twenty-three participants had strong STEM identities when entering college as a declared STEM major. When asked how students decided on their majors in the first place, most of the participants pointed to their competence and interest in various STEM content as sources of evidence in the decision-making process. They had demonstrated competence through the courses they took during their secondary education and had the ability to perform well in their various STEM classes. Many earned high grades in those courses and mostly enjoyed the STEM classes they took: "In high school, I really liked chemistry, like general chemistry, chem AP, chem Honors…. I really enjoyed it. I liked the problem-solving process of it." said Hannah (SN, female, Asian) who started college as a chemistry major. "I didn't exactly enjoy physics as much as I was anticipating… but I was good at it… In terms of math I was pretty involved in whatever math opportunities my school had to offer… So I really liked math." said Joshua (SS, male, Asian/Pacific Islander) who started as a biomedical engineering major.

Our participants recognized themselves to be "a STEM person" and they had recognition by meaningful others, which were also connected to their initial decision making. Angela (SN, female, Pacific Islander) expressed with conviction how her competence was high; she was recognized by her teachers, and felt connected with her teachers: "I did really well in high school, and I had a really good GPA, loved all my teachers, my teachers loved me, especially my STEM classes." This informed assumptions for college, "I assumed it would be the same thing. Oh, you're good at it [STEM] in high school, and you're gonna be good at it in college."

Students articulated how their environment contributed to their expectations and ultimately their STEM identities. Several students described coming "from family engineers" and engineering being "kind of what my family does" (Calvin, SS-adj, male, half Asian half White), or pointing out "also my dad, he did his undergrad in mechanical engineering, so that may have played a part in that [my initial choice]" (Jared, SS-adj, male, White). Jared was also the president of his high school's robotics club, suggesting that his beliefs were confirmed by his performance, which was also connected to what he felt was his family culture. Addie (SS-adj, female, White) definitively spoke about her growing up near Silicon Valley in California and how everyone at her high school was expected to pursue STEM majors. These findings align with previous research that students who left STEM were not different in terms of competence, interest, and performance from those who stayed in STEM at the beginning of college (Seymour & Hewitt, 1997; Shedlosky-Shoemaker & Fautch, 2015).

Among our interviewed participants there was one student who described herself being "pretty average" in STEM throughout high school. Lily (SS, female, Asian) actually described herself as an "English gal" and "the humanities person." She elaborated later in the interview that her reasons for becoming a STEM major initially was because her parents would not financially support her if she pursued a social science or humanities degree. The two remaining students who did not demonstrate a strong STEM identity did not share enough information to determine the strength of their STEM identities.

College as Context for STEM Identity Development College was a new context with new experiences that contributed to shaping participants' STEM identities. Some experiences were related to instructor pedagogy or course structure. Jared

(SS-adj, male, White) described how one of his physics professors came to class, opened the textbook, read silently to herself/himself, and then proceeded to lecture by talking and writing on the board. Some students perceived that prepared labs were poorly constructed. Daniel (SS, male, White) felt that his lab instructor "was just as lost as we were." Some students perceived instructor attitudes and behaviors negatively impacting their sense of belonging in majors and programs. However, unlike previous research, poor instructional quality was not universal across our campus. Some of our participants described how their experiences in STEM were useful, interesting, and facilitated learning: "I had a professor in Chem 111A ... and really liked the way she taught.... I took Bio 208, which was Anatomy, and that professor made it exciting ... I felt they were good in what they were teaching" (Joshua, SS, male, Asian/Pacific Islander). The positive experiences seemed to contribute to students' self-esteem and facilitate positive STEM identity.

Class size came up as a structural issue related to pedagogy and tutoring, and seemed to function as a barrier for students to be recognized. The large class structure, as many of STEM courses were, created environments that were very teacher-centered with little or no student interaction. The large class sizes for introductory courses were cited by several students as a factor in their decision to switch majors because they exacerbated student feelings of isolation: "The 100 Engineering class was a lecture course with 100 people and you're just kind of in the back.... It's like 'you know you're one in 100', then I was like 'okay, I go to class, I do the work.'" (Calvin, SS-adj, male, half Asian half White). These large courses were described as a place where little (if any) active learning or interactive pedagogy was implemented: "I understand because it's like a big lecture hall, and it is hard to really help students one-on-one." (Jacee, SS-adj, female, Asian). Class size also impeded on opportunities to connect with peers. Carl (SN, male, Asian/Pacific Islander) shared, "I feel in general that the [STEM] field can be lonely because people are just studying all the time, and there isn't [initially an] expectation for you to study with other people." These student experiences suggested challenges to being recognized as a STEM person by others around them.

Peers were important parts of the college context. Participants who had poor-quality science training and education prior to college described how they received support from peers or tutors in college and were able to be successful: "I didn't have a good science background. ... I was doing well in math, but physics and chem were *so* difficult. But I just needed that foundation and those people and those friends who actually pushed me through it." (Danae, SS, female, Asian). Whereas in contrast, Angela (SN, female, Pacific Islander) found "the biggest challenge for me was seeing all of my peers knowing what they were doing and me just feeling lost. And that's what really pushed me out.... It was so frustrating." So peers provided support and functioned as a norm with which to compare one's competence, performance, and interest.

Unexpectedly, family was a significant factor shaping student STEM identity during college. Parental support/pressure appeared as a big factor in students' choice of STEM majors and career pursuit, despite parent(s) and students themselves often not knowing what scientists and engineers did in their jobs. Sofia (SN, female, Asian) reported that her mom "always pushed me into going into something in the medical field, and specifically science, which I was interested in science." Other students described feeling obligated to work toward a STEM major even though that was not their passion. They later ended up switching and then breaking the news to their parents—which caused some stress. Angela (SN, female, Pacific Islander) said "I called my mom [after switching], and my mom just didn't accept it. She was, 'Okay, why didn't you do something else that can make you more money…. What do you think you're gonna do with a psychology degree?" Regardless of the process, most students found out that their parents really desired for their children to be happy, and that a STEM major was not a requirement in the end.

Challenged STEM Identity in College Against this backdrop of college, many students recalled experiences that challenged their STEM identity, particularly their feelings of competence and sustained interest in their respective STEM fields. One-third of the students spoke of wanting to apply their learning in some way, to be more actively engaged in class, to link class concepts to their lives and future careers; but felt they did not have an opportunity to do that in their first-chosen STEM majors. They described how they simply came to class, tried to keep up with a frantic pace of instruction, and ended up learning more procedurally rather than conceptually. Jacee (SS-adj, female, Asian) said, "Yeah… those problems. I understood how to do it. Just like step one, step two. But then I think overall, I just, I didn't get the conceptual part." Students identified failures and the emotions connected to those experiences contributed to their re-evaluation of being STEM majors (see Table 13.1 for switched majors). For some, the lack of understanding likely translated into a poor grade. Chloe (SS, female, Latinx) shared, "I got my first C ever here, and that killed me. It's actually depressing." This was not for lack of trying. Many of our participants discussed doing the practice problems to prepare for tests and seeking help both from their instructors as well as the tutoring services offered on campus. Despite these efforts, for some, the B at the end of the semester was "not good for me. It's just below average for me," (Kirsty, SS-adj, female, White) and the evidence of perceived poor competence challenged their positive STEM identity.

Challenges to STEM identity were experienced even among our participants who switched to another STEM major, specifically in the domain of interest. Joshua (SS, male, Pacific Islander) explored other engineering disciplines and found them to also not be of interest: "I felt those little small fundamental things of engineering that are supposed to entice you, they just didn't do the job for me." Jeremy (SS-adj, male, Asian) said, "I just couldn't see myself working on that kind of material for the rest of my life. It was just something that wasn't interesting to me." Importantly, we also found that not all challenges were negative. Lily (SS, female, Asian), our one participant who initially considered herself an "English gal", took an accelerated introductory chemistry course over the summer. She shared,

> Although it was stressful and extremely high-paced, it showed me not only kind of my own potential but it reminded me of that even if you can gauge your potential and figure out your strengths and weaknesses, but don't limit yourself to that (Lily).

13.4.2 Perceptions of STEM Disciplines and Switching Majors

Our second critical question was how STEM switchers perceived the four disciplines of STEM as interdependent and how their perceptions related to their choices of switching. Students' explanations of how they viewed the connections among STEM disciplines emerged into two groups. One group of students (fourteen of them) demonstrated their understanding of how STEM disciplines (Science, Technology, Engineering and Mathematics) were interrelated. Our team called this group having *sophisticated ideas* of STEM and these students often had more clearly articulated career goals. The second group of students (nine of them) demonstrated *superficial ideas* of what STEM was and how STEM disciplines were interrelated. Such as, many students in this category commented on their idea of STEM as being "hard and difficult" or as a subject for "smart people" or something that helped you make a lot of money. Summarized in Table 13.2 are the names/numbers of students who were categorized in two groups (sophisticated vs. superficial views of interdependence of STEM disciplines) and the type of majors they switched to and their final career goals.

Table 13.2 Understanding of interrelatedness of STEM and students' major switching patterns

Major switching	Sophisticated understanding		Superficial understanding		Total
	Name	Career goal	Name	Career goal	
STEM to STEM	Kirsty	Math teacher & considering other options	Danae	Environmental science in the fields of marine life*	**8**
	Kelsey	Math education			
	Lily	Pharmacy school			
	Chloe	Dentistry			
	Gaby	Engineering related*			
	Daniel	Biochem field*			
	Joshua	Pharmacy school			
STEM to STEM-Adjacent	Jeremy	Cybersecurity management	Addie	Not clear	**7**
	Calvin	Industrial designer	Jacee	Maybe work at a hospital field	
			Alicia	IT work in a hospital	
			Sarah	Nursing school	
			Jared	Economics graduate school*	

(continued)

Table 13.2 (continued)

Major switching	Sophisticated understanding		Superficial understanding		Total
	Name	Career goal	Name	Career goal	
STEM to Non-STEM	Mary	Hum.Res(HR), marriage or high school counseling	Angela	Clinical psychologist	**8**
	Hannah	Recreational therapy	Chris	Social work	
	Sofia	Not clear	Marco	Construction or working for a big company*	
	Tonya	Film industry			
	Carl	Divinity school			
Total	**14**		**9**		**23**

*not a specific career goal

Sophisticated Understanding of STEM Among the first group of students who demonstrated a sophisticated understanding of STEM, two mentioned STEM disciplines as a way of thinking: "You have to understand a certain level of math, physics, and chemistry to get around [STEM] ... Someone with more technical, science-oriented thinking, I think they definitely fall into this field, STEM." (Calvin, SS-adj, male, half Asian half White); "If I were to compare how I looked before college and after college into STEM, I would say that my way of thinking has changed." (Joshua, SS, male, Asian/Pacific Islander).

The majority of students in this category showed their understanding of the interdependence of STEM disciplines as they described the similarities or differences between distinctive subjects based on their STEM learning experiences. For instance, Gaby (SS, female, Latinx/Indian) explained how mathematics and physics were interdependent to each other from her own experiences: "My comprehension of Calc 2 helped me in my physics but that wasn't until later in the course when I was actually understanding what we were doing." In a similar way, Chloe (SS, female, Latinx) perceived STEM subjects as related based on her experience that tutors in STEM majors at the tutoring center were able to help students who were in any STEM majors:

> I think it [STEM] is very generalized, but I see why it is what it is like.... at the tutoring center, I've gotten to see that, they are tutors, they're math majors, but they can help you in other classes like chemistry. Yeah. I mean, it works! (Chloe).

Chloe also reported switching from biochemistry to microbiology "was quick and easy" and she made the move because she learned that microbiology aligned better with her career goals of pursuing dentistry school.

Chloe was one of the nine students who viewed STEM disciplines as interdependent and switched to STEM or STEM-adjacent majors. These participants also either started college with clear STEM or STEM-adjacent career goals or established those career goals quickly in their first year or two of college. Another interesting pattern we discovered was that all but one student who switched from their initial STEM majors to another STEM majors demonstrated a more sophisticated understanding of STEM.

Experiencing STEM disciplines as interrelated did not always result in positive outcomes. Learning more about the interdependence of STEM disciplines pushed some students to leave STEM entirely. Mary (SN, female, White) learned that mathematics played a role in learning chemistry, which in turn played a role in learning biology. So when she struggled with chemistry due to the heavy emphasis on mathematics, she switched from biology to psychology:

> I didn't realize that chemistry would have a lot of math in it because when I took it in high school, it wasn't really all about math. I didn't really relate math to chemistry a lot until I got into college…. They [advisor] told me that chemistry is a HUGE part of biology, if you can't really do that chemistry part, you might wanna switch your major…. I felt like after that chemistry class, 'I kind of wanted to branch out from STEM' at that point (Mary).

Five students, including Mary, with sophisticated understandings of STEM who switched to non-STEM majors either did not have clear career goals yet or pursued non-STEM careers. Carl (SN, male, Asian/Pacific Islander) shared that he became Christian during his senior year of high school, and in college his faith grew. He realized he no longer wanted to be a medical doctor, which he wanted to become because they make a lot of money. Instead, Carl said he wanted to become a pastor. He found his faith to be more important to him and found being a communication major supported his career pursuits.

Superficial Understanding of STEM The second group of nine students demonstrated a superficial understanding of STEM. Some of them recognized the similarities of STEM subjects only by the heavy workload: "One of the other things I would kind of associate with that [STEM] would be just how hard you have to work, the time that goes to put into." (Addie, SS-adj, female, White); "I think that they [STEM] are relatively similar…. they make STEM seem more like… 'oh, only the smart kids do STEM.'" (Angela, SN, female, Pacific Islander); "I see the engineers work 24/7 studying and don't have a life. The other science majors did study a lot and they were very dedicated." (Sarah, SS-adj, female, Latinx/Hispanic).

When these students struggled in STEM courses, they seemed to have sought out other meaningful personal values to help guide their college paths. Sarah (SS-adj, female, Latinx) talked about always wanting to work in the healthcare field. In high school she thought of becoming a physician assistant and was encouraged to pursue biology. Sarah received a D in her math course as a biology major in the first semester of college and gave up her major after that semester with limited STEM experiences: "I still felt so bad about that math class, so I was like maybe I can't do this…. So I kind of disregarded that college anymore. I didn't think it was an option for me." She found her way to becoming a health science major through advice from peers and was pursuing a nursing program after graduation. Thus, with the goal of being in the healthcare field, Sarah changed the profession she wanted to pursue.

Others in the same category demonstrated very naïve ideas of STEM as a means to "make a lot of money" or "get a job": "after graduation with just an undergraduate [STEM] degree, you can go to a company like Boeing and start getting paid around 65,000 or 50,000 just somewhere in that nice starting range." (Jared. SS-adj, male, White); "it was always like, 'Oh, be a good engineer, you make a lot of money' that way" (Marco, SN, male, Latinx); I also viewed it [STEM] as like, 'Oh, okay, STEM, I'm pretty sure I can find a job after.'" (Jacee, SS-adj, female, Asian). These students also viewed STEM only by its practicality without mentioning interdependence of each STEM discipline. Marco (SN, male, Latinx) who initially chose a mechanical engineering major realized that the high earning power of an engineering degree was not his life goal. He switched to consumer affairs and was uncertain about his future at the time of the interview. Jared (SS-adj, male, White) who also started in mechanical engineering with the same reason as Marco found economics-related careers were more promising for high earnings.

Students in this second group did not show clear patterns with regards to where they switched. Three of them did not have clear career goals as they described career ideas that were disparate from one another or described with hesitation. Four of them had practical reasons for switching that were related to their current GPA, expectations surrounding the prerequisites, or concern for delayed graduation.

13.5 Discussion and Implications

In the current study we examined the experiences of students who started college as a declared-STEM major and ended up switching to another major, using a STEM identity framework. As shown, in general students recalled having positive science and math experiences (a smaller number had experiences with engineering) in high school that led them to pursue STEM majors in college. They discussed many challenges and concerns they encountered while taking STEM courses in college. Those experiences challenged their STEM identities by making students question their past competence, interest, and belonging to rethink whether their initial STEM majors were the right place for them. Feelings and perceptions of failure functioned to push students away from STEM studies, creating an avoidant STEM identity and contributing to their decisions of switching out of the initial STEM majors. Our results echoed other research findings that identity plays a critical role in students' initial choice of study as well as whether they stay or leave STEM (e.g., Holmegaard et al., 2014).

Another result of this work is regarding the perceptions of STEM by STEM switchers and the relation between their perceptions and the choice of switching majors. Our findings revealed that the participants perceived the interdependence of STEM disciplines at different levels and their understanding of STEM was related to their choices of switching majors to some extent. We observed more often students who showed sophisticated understanding of interdependence of STEM moved to other STEM or STEM-adjacent majors after switching out of initial STEM majors. We questioned whether those students tried to keep their initial STEM identity intact as they switched to other or similar STEM fields. It is worthwhile to note that the term 'sophisticated understanding' does not mean students understood the integration of all STEM disciplines from epistemological and ontological underpinnings. Rather, students understood, based on their experiences, how distinctive STEM courses (e.g., calculus vs. physics) had an impact on their performance within the realm of STEM. This was not surprising due to the fact that students experienced STEM courses in compartmentalized disciplines. Students who described STEM as one discipline in and of itself for utilitarian purposes (e.g., money or job) tended to switch to non-STEM or STEM-adjacent majors when they did not have clear career goals.

A noteworthy point in the study is a students' career goal or image (i.e., helping people) played a pivotal role for all switchers in connecting their STEM identity narratives with their perceptions of STEM. The students who switched to non-STEM majors in spite of their initial strong STEM identity and interests did so partially because they signed up to be a STEM major without clear STEM career goals. As the students found out what career choices were available, they re-evaluated whether following the initial STEM majors would allow or limit the possibilities of who they might become. Thus, students' thoughts concerning their future jobs and careers could contribute to and conflict with the students' STEM identities. These career-related transitions and choices are defined as career development and it happens over a lifespan (Belser et al., 2018). As we considered ever-changing students' STEM identities in relation with their career goals as a career development process, we argue that students' STEM career goals should be included as a dimension of STEM identity. This is in contrast to research that looks at science identity and science career choices as two separate factors that have a strong relationship (e.g., Hazari et al., 2010). Furthermore, we recommend students explore different STEM careers while they are in high school, so they have a better understanding of STEM majors before entering college.

The college context was rich with experiences that discouraged and also encouraged student STEM pursuit. Our participants had and have strong work ethics and desire to persist. For some challenging courses and content material were opportunities to build self-confidence. For others who sought help, simply telling them to "study longer" or "study more" was not helpful. Quality of instruction was a key factor in reshaping students' STEM identity and switching decisions. There was also a need for application closely tied to learning of content and active engagement in class. Our students found introductory courses taught didactically in large lecture halls full of students challenging, consistent with current research on undergraduate STEM courses (Stains et al., 2018). This is critical because when students struggled, it was often in their prerequisite courses as shown in the previous studies (Crisp, Nora, & Taggart, 2009; Gasiewski et al., 2012; Mervis, 2010). Often, those prerequisite courses were perceived as being unrelated to the topic of study. We wondered whether student persistence in STEM could be changed through explicitly making connections between their prerequisite courses and their intended disciplines. Nonetheless, our students reported experiencing good STEM classes, and we considered this a reflection of efforts made to improve STEM teaching. Faculty need to continue efforts to change how these courses are taught as actively wrestling with ideas allow for deeper learning (Baxter & Gray, 2001; Wright, 2011) because one-semester courses have an impact on STEM students' persistence (Watkins & Mazur, 2013).

There are some limitations to note. As a qualitative interview study, our findings were not intended to be generalizable but generate significant knowledge in its own right, as a single sample case study can provide valuable insight (Boddy, 2016). However, we were limited in relying on students self-selecting to participate. Even with the help of the advising office in getting our recruitment information out to as many students as possible, we expect our sample is missing the voices and experiences of various students across STEM majors. Future research utilizing purposeful sampling techniques are needed to get a fuller picture of diverse students' STEM switching experiences, such as racial-ethnic minority STEM students, considering the context of the university as a minority-serving institute. The participants in this study were all from the United States and therefore international generalizations cannot necessarily be made as well. Thus, it would be interesting to seek out international students or to interview students in other countries in order to have a broader understanding of students' experiences in STEM majors. Students' ethnicities as well as their cultural experiences in different countries may impact students' perception of themselves and their experiences in STEM. Another limitation is regarding data collection. The data only relied on students' recollections of their experiences. Additionally, given that the research team was made up of all women, male students may have not felt as comfortable and we cannot know if we missed opportunities to probe or follow-up. Three of the research team are Asians and one is White. Therefore, it is difficult to determine how comfortable the Latinx participants were in the interview process and whether their unique experiences were fully captured and understood by the research team. Future research will benefit from incorporating data from instructors, advisors, and others who interact with the students and having more diverse research teams that can better collect and analyze data from diverse participants.

There are several noteworthy contextual aspects regarding the research site. The university is an HSI and AANAPISI. We questioned whether for some Latinx and Asian American, particularly Pacific Islander students, they were able to be successful in STEM or STEM-adjacent fields because they found places on campus outside of their majors where there were others who looked like them, helping their sense of belonging (see Freeman, Anderman, & Jensen, 2007 for class versus campus-level sense of belonging). Another contextual aspect less represented in the research literature on college students is the role of parents in the choice of STEM majors and STEM career pursuit. The university is a regional, public school and many students commute from home. Several students described how their parents voiced expectations, suggested refusing to pay tuition to pursue a non-STEM major, and later expressed holding onto hope that the student would still continue to pursue the STEM career. Students also voiced not wanting to disappoint their parents. So, although college students are legally adults, they still seek guidance from their parents. Lastly, in contrast with Maltese and Cooper (2017) who reported females in STEM fields not feeling supported by their parents, our study did not find clear gender differences in parental support. Future research would benefit from examining the role of parents in supporting or discouraging college students' STEM efforts.

13.6 Concluding Thoughts

Students switched to find new majors that aligned with their career goals with their understanding of STEM disciplines and their negotiated STEM identities. In many cases, our students expressed their desire to continuously use STEM knowledge and skills to make important contributions in society when they switched to other majors. Thus, we question whether lumping those students who switched out from initial STEM majors into one big umbrella such as STEM switchers is appropriate and even whether retaining all STEM entrants be the goal.

Given that there are differences across STEM subfields and the wide variety of possible trajectories in and out of the realm of STEM, we propose the metaphor of multiple lanes in investigating the STEM switching phenomenon. It allows researchers to explore lived experiences of students who may stay on the initially entered lanes to earn STEM degrees, who may switch to other STEM or STEM-related lanes, or who may totally exit the STEM lanes. For instance, we learned that the stories of exiting STEM lanes were not always the same. Angela (SN, Female, Pacific Islander) switched out from her initial STEM major due to the negative experience: "I remember the semester I decided to change was because of this professor, the very first day of class, was telling us, 'Oh, we're gonna weed out the people that shouldn't even be in here.'". Hannah (SN, Female, Asian), who would like to become a recreational therapist, left STEM because the original STEM major was not meeting her expectations and she found another discipline that was more interested in: "I like helping people directly, like interacting directly with people. And I like healthcare, the idea of healthcare professions, too.". These two students are simply counted as STEM switchers. However, listening to individuals' experiences let us reevaluate whether exiting the STEM lanes was problematic or not. In Angela's case, yes, higher education should be concerned about their STEM instruction or programs. Meanwhile, Hannah's case should be considered "a positive outcome, because it is part of the natural process of exploration and discovery in college" (National Academies of Sciences, Engineering, and Medicine, 2016, p. 4).

We acknowledge the metaphor of multiple STEM lanes raises questions of how many lanes are needed to explain the STEM switching phenomenon, in turn, how STEM subfields should be recognized. Moreover, there is still confusion where we would place other fields that are out of traditional STEM boundaries but require STEM knowledge and skills. Nonetheless, the metaphor allows researchers to pay attention to the diverse experiences and reasons college students have when switching out from initial STEM majors while considering their career choices and STEM identity negotiation. Students' interest, competence, performance, recognition, career goals, family expectations, and understanding of disciplines are all significant in both their current and future STEM pursuits. The multiple STEM lane metaphor captures how students experience these components as well as how they may differently weigh them at different times and in different ways, which in turn can help researchers and educators gain a better understanding of the STEM switching phenomenon.

Appendix: Interview Protocol

When did you start the university? When did you graduate? Are you still on campus?

What major did you start at the university? What major did you switch to? When did you switch?

Tell me about the decision-making process for choosing the initial major.

- Who and what influenced your decision?
- Tell me more about your high school STEM learning experiences.
- What was the image you had (or you think other people had) of STEM majors and STEM careers?
- What was the image you had of your major specifically?

Tell me about your experiences as a STEM major student.

- What STEM courses did you take before switching your major?
- Could you identify successes/challenges that you faced in taking STEM courses?
- What strategies did you use to be successful in your major?
- What sorts of support did you have to be successful? (both formal and informal)
- What did you think about your STEM instructors? STEM classmates?
- What sort of difficulties did you have? Were any related to your background?

Tell me about the decision-making process for changing your major.

- How did you pick the major to switch to? What did you consider? What did you think about?
- What were your feelings about switching?

Tell me about your experiences in your new majors.

- Did you consider another similar STEM major? What made you (not) pick any of those similar STEM majors?
- What kinds of similarities/differences did you notice in different STEM fields?
- In your own words, what are some differences in the nature of the STEM vs. non-STEM fields?
- Did you notice any differences between STEM vs. non-STEM courses? (e.g., level/types of difficulty, professors, classmates, support system)

Tell me about your career expectations or experiences.

- Are you working now?
- Could you describe changes in career plans related to switching? (consequences of, and feelings, about career shift)
- What's the image you have (or you think other people have) of STEM majors and STEM careers now?
- What's the image you have of this switched major specifically?

Please identify your ethnicity. What criteria do you use to claim your ethnicity?

References

Akerson, V., Burgess, A., Gerber, A., Guo, M., Khan, T. A., & Newman, S. (2018). Disentangling the meaning of STEM: Implications for science education and science teacher education. *Journal of Science Teacher Education, 29*(1), 1–8.

Aschbacher, P. R., Li, E., & Roth, E. J. (2010). Is science me? High school students' identities, participation and aspirations in science, engineering, and medicine. *Journal of Research in Science Teaching, 47*(5), 564–582.

Astorne-Figari, C., & Speer, J. D. (2018). Drop out, switch majors, or persist? The contrasting gender gaps. *Economics Letters, 164*, 82–85.

Atkins, P. (2018). *Atkins' Physical Chemistry*. Oxford: Oxford University Press.

Baxter, S., & Gray, C. (2001). The application of student-centered learning approaches to clinical education. *International Journal of Language & Communication Disorders: Supplement, 36*, 396–400.

Belser, C. T., Shillingford, M., Daire, A. P., Prescod, D. J., & Dagley, M. A. (2018). Factors influencing undergraduate student retention in STEM majors: Career development, math ability, and demographics. *Professional Counselor, 8*(3), 262–276.

Boddy, C. R. (2016). Sample size for qualitative research. *Qualitative Market Research, 19*(4), 426–432. https://doi.org/10.1108/QMR-06-2016-0053

Braun, V., & Clark, V. (2006). Using thematic analysis in psychology. *Qualitative Research in Psychology, 3*(2), 77–101.

Breiner, J. M., Johnson, C. C., Harkness, S. S., & Koehler, C. M. (2012). What is STEM? A discussion about conceptions of STEM in education and partnerships. *School Science and Mathematics, 112*(1), 3–11.

Byars-Winston, A. (2014). Toward a framework for multicultural STEM-focused career interventions. *The Career Development Quarterly, 62*(4), 340–357.

Callister, W. D., & Rethwisch, D. G. (2014). *Materials science and engineering: An introduction*. New York: Wiley.

Carlone, H. B., & Johnson, A. (2007). Understanding the science experiences of successful women of color: Science identity as an analytic lens. *Journal of Research in Science Teaching, 44*(8), 1187–1218.

Chen, X. (2013). *STEM attrition: College students' paths into and out of STEM fields* (NCES 2014-001). Washington, DC: National Center for Education Statistics, Institute of Education Sciences, U.S. Department of Education.

Chesky, N. Z., & Wolfmeyer, M. R. (2015). *Philosophy of STEM education: A critical investigation*. New York: Palgrave Pivot.

Crisp, G., Nora, A., & Taggart, A. (2009). Student characteristics, pre-college, college, and environmental factors as predictors of majoring in and earning a STEM degree: An analysis of students attending a Hispanic serving institution. *American Educational Research Journal, 46*(4), 924–942. https://doi.org/10.3102/0002831209349460

Denzin, N. K. (2017). *The research act: A theoretical introduction to sociological methods*. New Brunswick, NJ: AldineTransaction.

Dogan, B., & Robin, B. (2015). Technology's Role in STEM Education and the STEM SOS Model. In *A Practice-based Model of STEM Teaching* (pp. 77–94). Rotterdam: Sense Publishers. https://doi.org/10.1007/978-94-6300-019-2_6

Fayer, S., Lacey, A., & Watson, A. (2017). STEM occupations: Past, present, and future. *Spotlight on Statistics*. Retrieved from https://www.bls.gov/spotlight/2017/science-technology-engineering-and-mathematics-stem-occupations-past-present-and-future/pdf/science-technology-engineering-and-mathematics-stem-occupations-past-present-and-future.pdf

Freeman, T. M., Anderman, L. H., & Jensen, J. M. (2007). Sense of belonging in college freshmen at the classroom and campus levels. *The Journal of Experimental Education, 75*(3), 203–220.

Gasiewski, J. A., Eagan, M. K., Garcia, G. A., Hurtado, S., & Chang, M. J. (2012). From gatekeeping to engagement: A multicontextual, mixed method study of student academic engagement

in introductory STEM courses. *Research in Higher Education, 53*(2), 229–261. https://doi.org/10.1007/s11162-011-9247-y
Glaser, B., & Strauss, A. (1967). *The discovery of grounded theory: Strategies for qualitative research.* Chicago: Aldine.
Green, A., & Sanderson, D. (2018). The roots of STEM achievement: An analysis of persistence and attainment in STEM majors. *The American Economist, 63*(1), 79–93.
Grier, J. M., & Johnston, C. C. (2009). An inquiry into the development of teacher identities in STEM career changers. *Journal of Science Teacher Education, 20*(1), 57–75.
Hazari, Z., Sonnert, G., Sadler, P. M., & Shanahan, M. C. (2010). Connecting high school physics experiences, outcome expectations, physics identity, and physics career choice: A gender study. *Journal of Research in Science Teaching, 47*(8), 978–1003.
Holmegaard, H. T., Madsen, L. M., & Ulriksen, L. (2014). To choose or not to choose science: Constructions of desirable identities among young people considering a STEM higher education programme. *International Journal of Science Education, 36*(2), 186–215.
Kim, A. Y., & Sinatra, G. M. (2018). Science identity development: an interactionist approach. *International journal of STEM education, 5*(1), 1–6.
Kim, A. Y., Sinatra, G. M., & Seyranian, V. (2018). Developing a STEM identity among young women: A social identity perspective. *Review of Educational Research, 88*(4), 589–625.
Lee, K. T., & Nason, R. A. (2013). The recruitment of STEM-talented students into teacher education programs. *International Journal of Engineering Education, 29*(4), 833–838.
Maltese, A. V., & Cooper, C. S. (2017). STEM Pathways: Do men and women differ in why they enter and exit? *AERA Open, 3*(3), 1–16. https://doi.org/10.1177/2332858417727276
Mark, S. L. (2017). A bit of both science and economics: A non-traditional STEM identity narrative. *Cultural Studies of Science Education, 13*, 983–1003.
McGee, E. O. (2016). Devalued Black and Latino racial identities: A by-product of STEM college culture? *American Educational Research Journal, 53*(6), 1626–1662.
Mervis, J. (2010). Better intro courses seen as key to reducing attrition of STEM majors. *Science, 330*(6002), 306. https://doi.org/10.1126/science.330.6002.306
Nadelson, L. S., McGuire, S. P., Davis, K. A., Farid, A., Hardy, K. K., Hsu, Y. C., et al. (2017). Am I a STEM professional? Documenting STEM student professional identity development. *Studies in Higher Education, 42*(4), 701–720.
National Academies of Sciences, Engineering, and Medicine. (2016). *Barriers and opportunities for 2-year and 4-year STEM degrees: Systemic change to support diverse student pathways.* Committee on barriers and opportunities in completing 2-year and 4-year STEM degrees. In S. Malcom & M. Feder (Eds.), *Board on Science Education, Division of Behavioral and Social Sciences and Education. Board on Higher Education and the Workforce, Policy and Global Affairs*. Washington, DC: The National Academies Press.
Nugent, G., Barker, B., Welch, G., Grandgenett, N., Wu, C., & Nelson, C. (2015). A model of factors contributing to STEM learning and career orientation. *International Journal of Science Education, 37*(7), 1067–1088. https://doi.org/10.1080/09500693.2015.1017863
Perez, T., Cromley, J. G., & Kaplan, A. (2014). The role of identity development, values, and costs in college STEM retention. *Journal of Educational Psychology, 106*(1), 315–329.
Peters-Burton, E. E. (2014). Is there a nature of STEM? *School Science and Mathematics., 114*(3), 99–101.
Polnariev, B. A., Jaafar, R., Hendrix, T., Morgan, H. P., Khethavath, P., & Idrissi, A. B. (2017). Nourishing STEM student success via a TEAM-Based Advisement Model. *International Journal of Higher Education, 6*(6), 31–43. https://doi.org/10.5430/ijhe.v6n6p31
Seymour, E., & Hewitt, N. M. (1997). *Talking about leaving: Why undergraduates leave the sciences*. Boulder, CO: Westview Press.
Shanahan, M. C. (2009). Identity in science learning. *Studies in Science Education, 45*(1), 43–64.
Shedlosky-Shoemaker, R., & Fautch, J. M. (2015). Who leaves, who stays? Psychological predictors of undergraduate chemistry students' persistence. *Journal of Chemical Education, 92*(3), 408–414. https://doi.org/10.1021/ed500571

Snyder, C., Oliveira, A. W., & Paska, L. M. (2013). STEM career changers' transformation into science teachers. *Journal of Science Teacher Education, 24*(4), 617–644.
Stains, M., Harshman, J., Barker, M. K., Chasteen, S. V., Cole, R., DeChenne-Peters, S. E., et al. (2018). Anatomy of STEM teaching in North American universities. *Science, 359*(6383), 1468–1470. https://doi.org/10.1126/science.aap8892
Strauss, A., & Corbin, J. (1990). *Basic of qualitative research: Grounded theory procedures and techniques* (2nd ed.). Newbury Park, CA: Sage.
Thompson, R., & Bolin, G. (2011). Indicators of success in STEM majors: A cohort study. *Journal of College Admission, 212*, 18–24.
Tro, N. J. (2011). *Chemistry: A molecular approach*. Boston: Pearson.
Watkins, J., & Mazur, E. (2013). Retaining students in science, technology, engineering, and mathematics (STEM) majors. *Journal of College Science Teaching, 42*(5), 36–41.
Williams, J. (2011). STEM education: Proceed with caution. *Design and Technology Education: An International Journal., 16*(1), 26–35.
Wilson, Z. S., Holmes, L., deGravelles, K., Sylvain, M. R., Batiste, L., Johnson, M., et al. (2012). Hierarchical mentoring: A transformative strategy for improving diversity and retention in undergraduate STEM disciplines. *Journal of Science Education and Technology, 21*(1), 148–156.
Wright, G. B. (2011). Student-centered learning in higher education. *International Journal of Teaching and Learning in Higher Education, 23*(3), 92–97.
Xu, Y. J. (2018). The experience and persistence of college students in STEM majors. *Journal of College Student Retention: Research, Theory & Practice., 19*(4), 413–432.
Xue, Y., & Larson, R. C. (2015). STEM crisis or STEM surplus? Yes and yes. *Monthly Labor Review, 2015*. https://doi.org/10.21916/mlr.2015.14

Youngjin Song is a Lecturer in the Department of Science Education at California State University, Long Beach. As a former high school chemistry teacher, she is interested in science teachers' professional growth through reflective practices. Her research also focuses on science learning and teaching for culturally and linguistically diverse learners. She previously worked at University of Norther Colorado as an Associate Professor.

Ann Y. Kim is an Assistant Professor in the Department of Human Development at California State University, Long Beach. Her research focuses on identity development in STEM contexts. She previously worked at an alternative high school near Washington DC.

Lisa M. Martin-Hansen is a Professor of Science Education at California State University, Long Beach and a former elementary and middle school science teacher. Her research focuses on pedagogical choice and how pedagogy impacts student success in science learning and teaching in various science education environments. She is a past-president of the Association for Science Teacher Education and current department chair of Science Education at CSULB.

Elaine Villanueva Bernal is a Lecturer in the Department of Chemistry and Biochemistry at California State University, Long Beach. Her research focuses on professional development of STEM K-12 and postsecondary faculty, specifically educational technology integration. She previously worked as a Failure Analyst in the Aerospace industry.

Reflection on Part I: Natures of the Disciplines that Make up STEM

Valarie L. Akerson, Angela Burgess, Alex Gerber, Meize Guo, Taukir Kahn, and Steven Newman

The chapters in this section provide an excellent overview of the natures of the individual disciplines that make up STEM, namely, the natures of science, technology, engineering, and mathematics. It is clear that these disciplines, while interrelated, are separate and distinct ways of knowing and kinds of knowledge. Because they are distinct ways of knowing, they are separate disciplines.

Full integration of all disciplines together cannot be made because they are different ways of knowing and kinds of knowledge, yet connections can be made across the disciplines. In other words, STEM includes the individual disciplines of science, technology, engineering, and mathematics in a way that is meaningful and showcases the interdependence of the fields. It is therefore our perspective that there is not, and never will be a "nature of STEM" because STEM is not a discipline in and of itself. It is a curricular approach that can be used to teach about the individual STEM disciplines, as well as how to make connections across disciplines. Of course, this kind of curricular approach that attempts to integrate, or at least draw connections within and among the disciplines, should be the subject of research. As noted in Chap. 1, prior research on integrated curricula indicates a "mixed bag" of success in student outcomes at best, and generally results in some of the disciplines being included more superficially than other disciplines, depending on the expertise and interest of the teacher.

The expertise and interest of the teacher brings up another challenge of a STEM curriculum, and that being difficulties with teacher preparation as STEM teachers. How can we prepare teachers to fully conceptualize all content in each discipline, as well as the meaningful connections to be made among and across all disciplines, including knowledge of the nature of those disciplines? Chapter 1 suggests a possibility of adding a STEM capstone course to teacher preparation programs, but

V. L. Akerson · A. Burgess · A. Gerber · M. Guo · T. Kahn · S. Newman
Indiana University, Bloomington, IN, USA

V. L. Akerson, G. A. Buck (eds.), *Critical Questions in STEM Education*, Contemporary Trends and Issues in Science Education 51,
https://doi.org/10.1007/978-3-030-57646-2

also points out that even with such a course, it would almost certainly be targeted toward whatever discipline was the teacher's expertise. For instance, if the teacher were a chemistry major that would be seen as the main discipline, and connections could be made to other disciplines, which would not likely receive full attention due to the teacher's focus on chemistry. It will take effort to make connections to other science disciplines, let alone to other STEM disciplines.

The chapters focus partially on defining their own disciplines as part of STEM literacy. Chapter 1 provides an overview of Nature of Science as part of STEM, and how it differs from other ways of knowing. Chapter 2 discusses STEM, and particularly technology, in terms of preparing persons for the workforce. The chapter describes that knowledge of technology can provide avenues to a workforce, and points out that conceptualizing nature of technology can help address issues of equity and access to technology. Chapter 3 also indicates a workforce need for engineering, and also discusses knowing about nature of engineering as part of STEM literacy. They caution against the "design process" that is used in many STEM curricula should be avoided becoming the "scientific method" of engineering. Indeed, they report that most people do not accurately conceptualize what engineers actually do. Teachers who do not have backgrounds in engineering may not be able to include engineering in the curriculum in a way that it actually is clear as another way of knowing, as well as be able to make meaningful connections to other disciplines of STEM.

While Nature of Engineering is a subject of current research, possibly because of its inclusion in the *Next Generation Science Standards*, Chap. 4 points out that Nature of Mathematics has not often been the subject of mathematics education research, but being part of STEM, it would be important to include nature of mathematics as part of mathematics research so it would be clear what kinds of ways of knowing make up all stem disciplines. Chapter 4 indicates that STEM could provide a venue for contemplating nature of mathematics as individuals consider defining the individual STEM disciplines and seek to make meaningful connections across the disciplines. In essence, contrasting nature of mathematics with the natures of the other disciplines may be a way to explore nature of mathematics within STEM, to help teachers in making connections to other disciplines.

In essence, the individual disciplines that make up STEM are not only different kinds of knowledge and are developed differently, but the natures of knowledge for each discipline is also different. Because of these differences it is clear that—as in all integrated instruction—it is not possible to fully integrate the STEM fields into one kind of knowledge, and therefore there is no nature of STEM mainly because STEM is not a separate discipline, but in fact, a compilation of four disciplines in a curricular approach.

In the following sections we share research regarding perceptions of STEM, as well as programs in STEM, that attempt to meaningfully combine the STEM disciplines. Part II shares research regarding teaching STEM in meaningful ways, and Part III shares research regarding perspectives about STEM.

Reflection on Part II: Research into the Teaching and Learning of STEM

Gayle A. Buck

The chapters in this section give us a glimpse into the critical questions science teacher educators are asking about STEM teaching and learning, as well as how they are exploring those questions. The various chapters illustrate research efforts involving digital video games, environmental STEM, project-based activities based in local industries, and more broadly, preparing pre-service and in-service teachers to develop STEM units for their own classrooms. Together, they provide us with an understanding of how science teacher educators are addressing questions regarding what it takes to teach, or prepare others to teach, in STEM-focused learning environments.

The common aspects across the chapters are the theoretical definitions of STEM and rationales for studying STEM teaching and learning. The theoretical definitions the authors provide in their chapters reside within the interdisciplinary aspects of the various STEM disciplines. This common understanding is, in part, evident in the fact that they answered our call for research studies that *include all individual disciplines of STEM (science, technology, engineering, and mathematics) in a way that is meaningful and showcases the interdependence of the fields*. They make reference to STEM as a meta-discipline, an independent combination, and an interdisciplinary approach. In regard to research foci, the authors explore topics or innovations that touch on the various disciplines within STEM. Although there are differences in terms of which of the STEM disciplines are emphasized; often with the others holding supporting roles such as technological and mathematical tools.

The rationales these authors provide for studying STEM teaching and learning are similar in that real-world applications, preparing essential and high- quality STEM workers, and college STEM readiness are referenced. Such similarity is also reflective in the frequent use of project-based, problem-based and authentic learning approaches. This is especially emphasized in Dobrin's chapter focusing on

G. A. Buck
Curriculum & Instruction, Indiana University, Bloomington, IN, USA

V. L. Akerson, G. A. Buck (eds.), *Critical Questions in STEM Education*, Contemporary Trends and Issues in Science Education 51,
https://doi.org/10.1007/978-3-030-57646-2

implementing a STEM-driven project-based learning model. Also, as with any educational reform discussion, the efforts described in these chapters reflect the educational science standards. In general, STEM teaching and learning is presented as an authentic twenty-first century educational idea reflecting a changing world and changing educational expectations. They commonly note, however, that STEM teaching and learning is hampered by teachers' lack of experience and training in an interdisciplinary manner. Yin takes this even further when she notes, "not only K-12 teachers; but also teacher educators who traditionally work in a single disciplinary field often lack the understanding of and experience with integrated STEM education".

Whereas, the chapters are similar in terms of theoretical definitions of STEM and rationales for studying STEM teaching and learning; they differ somewhat in the practical application of STEM and, subsequently, the implications of educational research on STEM teaching and learning. For example, Akaygun & Aslan-Tutak present a module for introducing integrative STEM education to preservice science and mathematics teachers. Their efforts focused on integrating chemistry and mathematics at the core, with technology and engineering tools and processes integrated into that core. Thus, they focused more heavily on science and mathematics. Burgess & Buck focused on integrating the four disciplines around the theme of Environment Science. They used the aspects of science, technology, engineering and mathematics used to explore an environmental topic. DeCoito & Briona had their preservice teachers develop digital video games to teach STEM concepts in biology, chemistry or physics. The teachers needed to use programming math, engineering and technology principles to develop the games. Thus, the content focus leans more heavily toward science. The games the teachers developed for their students, however, included three or four STEM subjects. Overall, all the chapters in this section focus on integrating the various disciplines in STEM, but there are often one or more disciplines that dominated.

Like the chapters in Part I, the chapters in this section reflect STEM as compilation of the four disciplines, through which interdisciplinary connections are made. Although the implications are interdisciplinary – the various individual disciplines of science, technology, engineering and mathematics are recognizable throughout. The teacher educators behind these interventions and processes share their attempts at overcoming the difficulties referenced in the first section; specifically requiring teachers, and teacher educators, with background and expertise in one of the disciplines to become STEM teachers.

Reflection on Part III: Critical Questions in STEM

Valarie L. Akerson and Meize Guo

In this section the research on STEM focused on how different populations view STEM, and therefore, how it is seen. For example, in the Sgro et al. chapter, the focus was on how STEM activities were portrayed for practitioners in teacher journals, in the Newman, et al. chapter the focus was on how schools portrayed themselves as STEM schools through their online presences, in the Madden et al. chapter interviews were conducted of elementary teachers to determine their perceptions of STEM, and in the Song, et al. chapter interviews of college students were used to determine not only their perceptions of STEM, but why some stayed STEM majors and why some switched majors. Perhaps not surprisingly, all researchers in this section used qualitative methods to explore their questions, which makes sense given they were exploring perceptions of STEM itself. It did not appear that the stakeholders held a definition of STEM as a meaningful interdependence among all disciplines that comprise STEM. In other words, there was a lack of perception that all individual disciplines of STEM should be included in ways that are meaningful and showcase the interdependence of the fields.

The chapters in this section reported the importance of STEM that has been recognized by different stakeholders. For instance, Sgro et al. and Newman et al. started from the groups who identified themselves as STEM groups, the STEM curriculum designers and STEM certificated schools, that have apparent practices in STEM education. In terms of STEM major switchers, Song et al. reported that although students might switch from a STEM major, they still value the STEM education and have a "strong work ethic and desire to persist" (p.22). Madden et al. report the results in the most direct way; their survey data shows that 97% of preservice teachers confirmed that STEM education is important.

In the cases of these chapters, it appears that individuals' interests influence them in their STEM enactment. In the case of the preservice teachers, the majority

V. L. Akerson · M. Guo
Indiana University, Bloomington, IN, USA

V. L. Akerson, G. A. Buck (eds.), *Critical Questions in STEM Education*, Contemporary Trends and Issues in Science Education 51,
https://doi.org/10.1007/978-3-030-57646-2

believed that STEM was important, and mostly for their students' futures. One imagines that those teachers would emphasize STEM in their instruction. In the case of the college students, oftentimes their pasts (including teachers and parents, and perceptions of self) influenced them to pursue a STEM path in college. However, if their perceptions of themselves as "STEM" people changed, or if they found stronger interests outside of STEM, they tended to change their majors. In the case of the STEM certified schools, the connection is similar to the preservice elementary teachers. It was apparent that schools went through a stringent application process to become certified, to support their students' STEM learning. Through their websites and social media they promoted ideas about STEM and how they were using STEM in their schools. In the case of the practitioner journals, the journals themselves sought to bring STEM lessons to the teachers so they could be enacted in classrooms. Despite the journals generally having a focus on one of the STEM disciplines, they sought to be cross/interdisciplinary and provide lessons and resources for teachers to provide STEM lessons to their students.

This focus on cross/interdisciplinarity in all chapters highlights the idea that STEM should be a meaningful integration across all STEM disciplines, in practice, it is generally one of those disciplines that comes to the forefront. For example, in the Sgro et al. chapter they suggest that in a STEM lesson, a chemistry teacher would teach chemistry, and then connect the other STEM disciplines of technology, engineering and mathematics. But still chemistry would be the main focus. Similarly, in the Newman et al. chapter, the STEM schools generally focused on mathematics and science, while making some connections to technology and engineering. Engineering was often seen as robotics in the portrayal of STEM by these STEM schools. The emphasis on STEM education and STEM careers really depend on background of the person, and also set up of the schools and universities. STEM education depends on the stakeholders and the contexts in which they find themselves.

In addition to highlighting the integrated character of STEM education; the chapter authors also discuss the challenges of STEM education. For example, Sgro et al. mentioned that different stakeholders, such as politicians, teachers, and parents, conceptualize STEM education in different ways due to their different goals. In the Newman et al. chapter, the authors found that there was no consistency of how to teach STEM nor what STEM means across grade levels. Song et al. also mentioned that pre-service teachers reported a gap between "willing to do STEM" and "less opportunity to do STEM education in practice." Also, the quality of instruction and future career opportunities influenced the college students' concerns and choices of major. Although STEM education is gaining popularity in the schools, curriculum design, and teacher education, for practitioners, there are still many barriers to effective STEM education.

Afterward

Dana L. Zeidler

STEM: Unrequited Dreams in a Material World

An array of dedicated and sincere scholars who have embraced STEM education, or at least aspects of STEM education, have crafted the chapters written for this volume so that it may be studied and advanced in sectors of the science education community. Let me be clear, I am not suggesting that all the authors advocate STEM education to the same degree, nor am I suggesting that all authors envision STEM that prioritizes it in a manner that usurps other arguably competitive goals in science education. Collectively, the authors have made clear that STEM is, in and of itself, not a discipline and success with STEM programs have been mixed. The chapters rightfully examine how, for example, the nature of science (NOS) may be realized as a way of knowing within STEM initiatives, how the nature of technology can possibly address equity issues related to access to information, how it may be used to beef up the pipeline in preparing students for the workplace, and the extent to which it is connected to scientific literacy. Missing from these, and related discussion throughout this volume, is the protagonist who might ask what some may perceive as curmudgeon-like questions. This is essentially the role I have embraced in the discussions of STEM education found in the deep, dark corners of the literature.

My objections to STEM-related initiatives have been argued in detail elsewhere and advance points ranging from positioning it as a deficit framework (Zeidler, 2016) to cautionary tales for humanity (Zeidler et al., 2016). Those objections are raised from the observation that STEM seems to be treated *sui generis* as an accepted social fact, one that is, in and of itself, desirable and worth pursuing even if it is a synthetic reality. This is precisely the distinction that Durkheim between 1898 and 1911 forewarned us about – that social facts or norms, desirable or not, tend to bond

D. L. Zeidler
Distinguished University Professor, University of South Florida, Tampa, FL, USA

V. L. Akerson, G. A. Buck (eds.), *Critical Questions in STEM Education*, Contemporary Trends and Issues in Science Education 51,
https://doi.org/10.1007/978-3-030-57646-2

a community together (Pocock, 1953), not unlike the science education community's embracement of STEM. The main problem at-hand is not the ubiquitous of STEM in our schools and in our scholarship, it is that many some undetermined number of supporters, likely more than not, have assumed that social fact means we ought to pursue it. Those that fall into that camp may have failed to ask the "open question" thereby committing a kind of naturalistic fallacy (Moore, 1903). It may be the case that STEM now seems to be part of our social fabric, yet we can reasonably ask, but is that good … is that desirable? To the extent that policy decisions in (science) education deal with the well-being of our students, such decisions are necessarily *moral* decisions, and deriving "ought" positions (STEM ought to be pursued) from statements about fact (STEM initiatives are currently being promoted and funded) is to commit that fallacy.

This is potentially problematic on several fronts. The most severe threat, however, is found in the dominant subjective hegemony created by the emphasis on training for the production of material goods for consumers in a material world. In such a world, emphasizing pipelines that lead to jobs in industry do so at the expense of sociocultural educational factors derived from the humanities. Cogs are overvalued at the expense of unrequited dreams, which lead to other creative endeavors. STEM education objectives ignore, at least as typically practiced, the formation of individual and collective consciousness. I don't argue that STEM education can not promote critical thinking; I do argue that there are qualitative differences, in-kind, between the types of critical thinking required to solve mechanical problems and the forms of critical thinking required to solve the kinds of ethical needs embedded in the fabric of worldly affairs.

Perhaps it comes down to asking a rather simple question about the purpose of teaching:

Do science teachers and science teacher educators have a moral imperative to provide the educative experiences necessary for the cultivation of character and justice? If the answer is "no," then the discussion ends here. If the answer is "well, yes," "maybe," or "let me think about that and get back to you," then we can proceed to advance a conversation about sociocultural perspectives of science education that becomes more inclusive and create opportunities for students to practice thinking creatively and fruitfully in everyday contexts that impact our family, community, societies, global networks of relationships, as well as our physical world.

It is easy to chalk such criticism about STEM education off as the ranting of the lunatic fringe (Roosevelt, 1913). But there are votaries found both with current STEM movements and those who view STEM as a kind of Trojan horse for neoliberal marketplace policies, and, in turn, advocate for the deconstruction of STEM in favor of more liberating approaches to science education (Bencze, Reiss, Sharma, & Weinstein, 2018). While it's appropriate to raise concerns about global economic competition among nations, it is certainly inappropriate and disingenuous to place those concerns above such issues as environmental sustainability, cultural dignity and compassion. These latter three examples are, of course, central to movements like socioscientific issues (SSI) social justice, socio-political action, ecojustice and the like. Other advocates of socio-cultural approaches (e.g., Evagorou, Nielsen, &

Dillon, 2020; Gough, 2014; Levinson, 2013) make similar arguments about the need to contextualize science within the human condition, and cautioning about the "power dynamics behind the push for STEM learning as an ideological discourse propagated by global networks of elite policy actors and enacted by non-elite policy actors at the school level" (Ellison & Allen, 2018, p. 267).

The arguments put forward for the contextualization of science have also recently been highlighted in (Bencze et al., in press; Kahn & Zeidler, 2016; Zeidler, Herman, & Sadler, 2019). In the absence of explicit SSI-type of approaches to science education pedagogy, tensions between more traditional essentialist expectations and the kind of creative interdisciplinary approaches beyond the scientific disciplinary connections within STEM that are associated with progressive practices become amplified. Such "mixed signal" pedagogical expectations are, no doubt, at least partially connected to movements like STEM, and are also tied to intense economic competitiveness that have prioritized the products of STEM over the cultivation of science-in-context (SinC) learning that entails critical thinking, reflective judgment, ethical considerations and moral reasoning. Bencze et al. (in press) provide an analysis of the arguments that extend SinC educative approaches like Socioscientific Issues, Socially Acute Questions, Science, Technology, Society, Environment, by examining the ontological, epistemological and pedagogical presuppositions of each that contribute to sociocultural robust conceptualizations of functional scientific literacy. While nuanced differences in these three SinC approaches exist, all are focused on problematizing reductionist STEM education initiatives through progressive efforts aimed at the development of critical consciousness. These efforts would also be value-added in the sense that they are consistent with the formation of a collective consciousness (Durkheim, 1893) and collective moral responsibility (Isaacs, 2011).

This pushback of the neoliberal features that characterize aspects of present STEM education is important from the sociological perspective of Durkheim in that both the expressions and limitations of science are bound by the collective representations of culture. Accordingly, regardless of the ontological status of reality, the collective representations of culture impact the extent to which rationality, scientific expression and knowledge are interpreted and conveyed. To the extent that the myopic focus on STEM educational initiatives represent a kind of null curriculum, the opportunity for excising important sociocultural contextual humanistic reservoirs of applied understanding (i.e., humanities, history, politics, etc.) as important element in the teaching and learning of science passes us by.

References

Bencze, J. L., Pouliot, C., Pedretti, E. Simonneaux, L., Simonneaux, J., & Zeidler, D.L. (in press). SAQ, SSI & STSE education: Defending and extending 'Science-in Context.' *Cultural Studies in Science Education*.

Bencze, L., Reiss, M., Sharma, A., & Weinstein, M. (2018). STEM education as 'Trojan horse': Deconstructed and reinvented for all. In L. Bryan & K. Tobin (Eds.), *Thirteen questions in science education* (pp. 69–87). New York: Peter Lang.

Durkheim, E. (1893). *The division of labour in society*. (W. D. Halls, Trans.). New York: Free Press.

Ellison, S., & Allen, B. (2018). Disruptive innovation, labor markets, and Big Valley STEM School: Network analysis in STEM education. *Cultural Studies of Science Education, 13*, 267–298.

Evagorou, E., Nielsen, J. A., & Dillon, J. (Eds.). (2020). *Science teacher education for responsible citizenship: Towards a pedagogy for relevance through socio-scientific issues* (pp. 3–6). New York/Dordrecht/Heidelberg/London: Springer.

Gough, A. (2014). STEM policy and science education: Scientistic curriculum and sociopolitical silences. *Cultural Studies of Science Education, 10*(2), 445–458.

Isaacs, T. (2011). *Moral responsibility in collective contexts*. Oxford: Oxford University Press.

Kahn, S., & Zeidler, D. L. (2016). Using our heads and HARTSS*: Developing perspective-taking skills for socioscientific reasoning (*Humanities, Arts, and Social Sciences). *Journal of Science Teacher Education, 27*(3), 261–281.

Levinson, R. (2013). Practice and theory of socio-scientific issues: An authentic model? *Studies in Science Education, 49*(10), 99–116.

Moore, G. E. (1903). *Principia ethica*. Cambridge: At the University Press.

Pocock, D.F. (1953). *Sociology and philosophy*. (D. F. Pocock, Trans.). London: Cohen and West. (Includes Durkheim's "Individual and Collective Representations" (1898), "The Determination of Moral Facts" (1906), and "Value Judgments and Judgments of Reality" (1911).

Roosevelt, T. (1913). *History as literature*. New York: Charles Scribner's Sons: Bartleby.com. Retrieved from: https://www.bartleby.com/56/11/html. 2/1/2020

Zeidler, D. L. (2016). STEM education: A deficit framework for the 21st century? A sociocultural socioscientific response. *Cultural Studies of Science Education, 11*(1), 11–26.

Zeidler, D. L., Herman, B. C., Clough, M. P., Olson, J. K., Kahn, S., & Newton, M. (2016). Humanitas emptor: Reconsidering recent trends and policy in science teacher education. *Journal of Science Teacher Education, 25*(5), 465–476.

Zeidler, D. L., Herman, B. C., & Sadler, T. D. (2019). New directions in socioscientific issues research. *Disciplinary and Interdisciplinary Science Education Research, 1*(11), 1–9. https://doi.org/10.1186/s43031-019-0008-7

The manufacturer's authorised representative in the EU is Springer Nature Customer Service Centre GmbH, Europaplatz 3, 69115 Heidelberg, Germany. If you have any concerns regarding our products, please contact ProductSafety@springernature.com

Printed and bound by CPI Group (UK) Ltd, Croydon, CR0 4YY
22/09/2026
02221061-0010